Eat, Cook, Grow

Eat, Cook, Grow

Mixing Human-Computer Interactions with Human-Food Interactions

Jaz Hee-jeong Choi, Marcus Foth, and Greg Hearn

The MIT Press
Cambridge, Massachusetts
London, England

MIT Press books may be purchased at special quantity discounts for business or sales promotional use. For information, please email special_sales@mitpress.mit.edu.

This book was set in Stone Serif Std by Toppan Best-set Premedia Limited, Hong Kong. Printed and bound in the United States of America.

Library of Congress Cataloging-in-Publication Data

Eat, cook, grow : mixing human-computer interactions with human-food
interactions / edited by Jaz Hee-jeong Choi, Marcus Foth, and Greg Hearn.
    pages   cm
Includes bibliographical references and index.
ISBN 978-0-262-02685-7 (hardcover : alk. paper)   1. Dinners and dining.   2. Agriculture.
3. Online social networks.   4. Food—Social aspects.   I. Choi, Jaz Hee-jeong, 1980- editor of compilation.   II. Foth, Marcus, editor of compilation.   III. Hearn, Greg, 1957- editor of compilation.
TX737.E28 2014
641.5'4—dc23
2013029585

10   9   8   7   6   5   4   3   2   1

Contents

Acknowledgments

This book has taken a long time to get here. As with any other works of mind, the book is a result of a colossal body of knowledge that has accumulated over time, some of which is difficult to pinpoint and to address accordingly. We thank the many thinkers who have actively voiced their thoughts that in turn advanced our understanding about the world, including the key themes of this book.

More directly, we owe a great deal of gratitude to people who have directly been involved in the *Eat, Cook, Grow* research project and this book. It simply would not have been possible without support from the following individuals and organizations: the *Eat, Cook, Grow* team including Shawn Lawson, Eli Blevis, Younghui Kim, Tad Hirsch, Robert Imre, Peter Lyle, and Geremy Farr-Wharton, and our partners Elizabeth Good at Queensland Health, Emma Kate-Rose and Robert Pekin at Food Connect, Peter Kearney at CityFood Growers, and John Meredith at James Street Cooking School.

We are also genuinely grateful to Doug Sery and people at The MIT Press for guiding us through the eventful path to publication, and to our colleagues around the world who kindly reviewed the chapters, including Nic Bidwell, Laura Forlano, Michelle Hall, Mariann Hardey, Jillian Hamilton, Ben Kraal, Eleftheria Lekakis, Conor Linehan, Nancy Odendaal, Jeni Paay, Gavin Sade, Christine Satchell, Ronald Schroeter, Mark Shepherd, and David Wright, among others. Natalie Collie has been nothing less than the best copyeditor one could wish for. And we would like to express our most sincere gratitude to the authors for sharing their valuable insights. Anne Galloway, Yvonne Rogers, Kenton O'Hara, Eli Blevis, and Charles Spence, thank you so much for your contributions, too.

Jaz, Marcus, and Greg

I am genuinely grateful to Marcus and Greg for being the most amazing companions in the journey of turning the *Eat, Cook, Grow* project and this book into reality; Mark Graham, Bernie Hogan, and Victoria Nash at the Oxford Internet Institute for their support in exploring new terrains of research in beautiful Oxford; Rob Comber and

Patrick Olivier for the same reason in wild wild Newcastle; Jean Burgess, Mandy Thomas, Genevieve Bell, and Adam Greenfield for magically appearing with light when I was drowning in the dark sea of intellectual and otherwise inanity; friends and colleagues in many different parts of the world for stimulating conversations and delightful food we shared; my family, especially my beautiful mother and Eli. I simply wouldn't be here without you. Thank you.

Jaz

My sincere thanks to Jaz who has been leading this program of research courageously and with such vision, creativity, and determination. It has been wonderful to witness how a simple idea over coffee years ago has now gained such momentum with a respectable body of work and international engagement driving it. I would also like to thank everyone at the Urban Informatics Research Lab at QUT who volunteered to function as sounding boards, reviewers, app testers, and dinner guests. I really appreciate the research culture of the lab and the supportive and constructive environment it creates.

Marcus

I would like to thank everyone involved in the project. It has been a fascinating and productive collaboration. I would especially like to acknowledge the creativity, hard work, and intellectual generosity of my co-conspirators, Jaz and Marcus.

Greg

Introduction

Jaz Hee-jeong Choi, Marcus Foth, and Greg Hearn

In 2008, we decided to collaborate on a research project because of a common concern reverberating among us. The concern was, and still is, that existing practices and systems of food do not ensure food security, but rather seriously threaten it, for most of the people in the world. Our aim was to find ways to encourage a major shift in the trajectories of food futures by bringing together our expertise in design, computing, urban informatics, media and communication, and social studies. Despite our common vision, we then had quite different ideas as to *how* we could realize it by navigating through the complex web of food across people, place, and technology. After three years of working together on the project called *Eat, Cook, Grow: Ubiquitous Technology for Sustainable Food Culture in the City* (http://www.urbaninformatics.net/projects/food), we still have varying ideas about many aspects of the topic. However, such divergence in thoughts comes as no surprise, given the pervasive and diverse nature of food discourse. Food is critical to our sustenance and, in its varied forms, is often an object of pleasure and desire with equally varied intensity. As such, while food may create personal moments of adversity or jouissance, it has also played an unquestionable role in shaping the broader political and social agendas throughout human history, and has co-evolved with the tangible and intangible facets of human society.

The World Food Programme's Hunger Map (http://cdn.wfp.org/hungermap) clearly indicates the strong correlation between food insecurity and civil unrest across the globe. In recent years, food security has reemerged as a major global concern. Its definition has changed over the years, too. In contrast to the earlier versions, which focused on the sustainability or equilibrium of the food system, the current definition by the Food and Agriculture Organization of the United Nations (FAO) emphasizes the importance of securing access for individuals, households, and communities to not only food but also sociocultural knowledge and practices around food: "a situation that exists when all people, at all times, have physical, social and economic access to sufficient, safe and nutritious food that meets their dietary needs and food preferences for an active and healthy life" (FAO 2001).

This definition reflects food's complex and scalable influence from macro level civil security to micro level visceral experiences. In fact, this ontological complexity has led to various rhetorical claims such as "Food is culture" (Montanari 2006) and "Tell me what you eat, and I will tell you what you are" (Brillat-Savarin 1999 [1949]). These have been used, in turn, to make various conflicting arguments in contemporary times across diverse domains: from diet and body image (or, at least, the marketing of such things) to grassroots food sustainability movements.

While such differences in opinion keep discussions more interesting for the broader scholarship as well as for the three of us, there is an acute need for a common thread in our necessarily multi-perspective approaches to understanding food if we are to generate actionable knowledge and henceforth move forward to secure our food futures. Among the three of us, we have come to at least one agreement, which in turn has allowed us to cook up this book together with contributions from colleagues who have equally interesting and diverse ideas about the topic. We all agree on the utmost significance of *people* as growers, cooks, and consumers of food and—most importantly—as active agents of food systems who are fundamentally influenced by the system but at the same time shape the system itself through their varied actions and decisions. Therefore, we place our hope in people and tactical change making that adds a bottom-up perspective to the conventional top-down agenda.

Further, we focus on *urban environments* as the place for change making for two main reasons. First, they are the fastest-growing form of human habitat across the world, with more than 60 percent of us expected to live in cities by 2030 (UNFPA 2007). The second reason concerns the consonant relationship between current pervasive technologies and urban spaces and formations. We note, however, that we are not attempting to frame our argument around the binary distinction between urban and rural as famously offered by, for example, Louis Wirth (1938) and others in the Chicago School. In fact, we find the growing urban–rural segregation as one of the greatest contributors to declining global food security. Our focus on urban environments is rather an acknowledgment of the nascent, intricate, and novel assemblages of opportunities for change presented by urban-technical affiliations, particularly at a level of individuals' everyday lives. We call here upon Saskia Sassen's concept of "cityness," which embraces complexity and multiscalar convergence of various forms of urbanity and acknowledges "that the intersection of differences produces something new, whether it is good or bad, and that this actually occurring intersection is consequential" (2005). This cityness creates incompleteness (Sassen 2011) or overspills (Choi and Greenfield 2009) that are messy (Bell and Dourish 2007) and which keep the city in a constant and necessarily transformative state that in turn keeps the city resilient.

Therefore, our collaboration aims to find ways to engage people in what could be considered quotidian practices of eating, cooking, and growing food and to make messes or overspills that shift the current configurations toward more environmentally sustainable, socially inclusive, and healthier futures. In our view, this is one of the most important questions that we as humans and scholars across all fields of research must ask at this point in history—when climate change has progressed from being a potential or imminent problem to a central and current situation that requires our immediate response at a global scale. Today we have new ways to know, connect, and live and to create responses in sustainable and democratic ways through the appropriate integration of technologies.

For this, we are inspired by undoubtedly the most significant and perhaps most mysterious turn in human history, which bears a curious degree of resemblance to our current state: the Neolithic Revolution. This was the transformative period when the tipping point was reached to make way for a fundamental transition from foraging to farming. As Michael Kimball (1998, 20) notes:

Considering the likelihood that anatomically modern humans, that is, *Homo sapiens sapiens*, have been around for at least 100,000 years (based on fossil evidence), the pace of that kind of change is remarkable. To put it in perspective: if we were to make a 100,000 year anatomically modern history equal to one 24 hour period, the transition to agriculture world-wide would have happened very late in the "day" and very swiftly—from about 9:30 to 10:45 PM.

There are varied speculations about what caused this rather late transition. In fact, there are at least thirty-eight distinct and competing causes for it, according to Ofer Bar-Yosef and Anna Belfer-Cohen (1992). While this transformation remains blurry to us, there has been a clear transformation in our thinking of what might have happened. Earlier studies in archeology and anthropology argued that the main causes were external, focusing on the economic model of technological advancement. That is, similarly to other forms of technologies such as the steam engine or the computer, "cultivating plants rather than foraging wild species is said to have raised the productivity of human labor, encouraging adoption of the new technology and allowing farming populations to expand" (Bowles 2011, 4760). Life expectancy did grow eventually, but evidence suggests that nomadic foragers had better productivity and health than an early generation of farmers; they had access to abundant food supplies; and they also had more leisure time. Marshall Sahlins famously called them "the original affluent society" (1972, see 1–39), in strong contrast to older ideas about a life of mere subsistence. Hunters and gatherers lived in a society that was shaped and advanced by human needs but also by *desires*. Given this, more recent studies (Lee 1968; Sahlins 1972) argue that the delay in adopting the agricultural way of life was not a surprise, but a natural progression engendered by two intersecting factors: first, critical changes in climatic conditions (Ashraf and Michalopoulos 2010) and second, endogenous and sociocultural factors, accompanied by chance encounters and playful experimentations (Price and Bar-Yosef 2011, 168; Sauer 1952) driven by *Homo sapiens* (or, using the term Johan Huizinga coined in 1938 to characterize humans as naturally playful beings, *Homo ludens*).

It is in this terrain that we see potential for most resilient and effective uses of *technologies* in urban environments toward optimistic food futures: through playful and human-centered engagement. We define "playfulness" per the concept proposed by Jaz Hee-jeong Choi (2010) in her research into urban networks. She argues that play is a voluntary, transient, and innovative experience in between the *pressure of control* and the *possibility of freedom*; this creates multiscalar messy overspills (Bell and Dourish 2007; Choi and Greenfield 2009), forming the core of cityness (Sassen 2005) for urban sustainability or resilience. We apply this concept to designing technologies that support making legible the current situation *as well as* articulating the possibilities for change, with a hope that the aggregation of such folding, unfolding, and refolding of ideas and situations—or contestations—would make a way to broader systemic change. This process requires careful planning and operationalization. A number of current trends in the technosocial sphere signal affirmatively to this hopeful notion. Key examples are the global bottom-up movements to challenge the once-walled garden of science, knowledge, and power, as readily observed in emerging hackerspaces, fab labs, Occupy movements, and DIY (Do-It-Yourself) / DIWO (Do-It-With-Others) initiatives across numerous thematic domains. Eloquently labeled as "read-write" (Greenfield 2010) and "open source" (Sassen 2011) by key thinkers in urban studies (as well as in digital and network technologies), the notion of urbanism engenders new spaces where people's voices can be made, heard, and lead to consequential changes to the city. Through these dialogic processes, people shape the city as the city shapes people.

This is precisely where we ground the latter half of this book's title. We deliberately use "human-food interactions" instead of, for example, "human-food experiences" to denote the *inter*connection between the self and food; food fundamentally influences the self and, at the same time, a person's actions also significantly influence—beyond individual food choices—the existing food and related systems. A number of recent projects exemplify this: for example, the Trash | Track[1] project by colleagues at the MIT sense*able* city lab and more recently, FoodMood,[2] a real-time twitter mapping of what people are eating. Digital fabrication (Malone and Lipson, 2007; Wei and Cheok 2012) and DIY food sciences (Ledford 2010; Wolinsky 2009) signal a new horizon of participatory food systems. The burgeoning interest in open sourcing in genomics (e.g., Arnquist 2009) also offers a glimpse at the future of perhaps more profound, though temporally demanding, developments in this terrain that may revolutionize science and medical research. However, these necessarily entail immense regulatory and ethical complexities.

We hope by now it is clear that we are not advocating for the use of technologies as a means of turning *back* to "nature" or "the ways of olden days" as the often-perceived "other" of contemporary urban living. Rather, we are arguing for the need to explore opportunities to create interfaces that help make legible potential uses toward healthy, socially inclusive, and environmentally sustainable food futures; and support people's engagement and collaborations in such processes of contestation, engendering new forms of urban networks and experiences. Furthermore, we are not proposing that technical means of framing and participating in the contestation are the only or the most effective way forward. We are simply pointing to a particular tendency that is inherent with potentially escalating capacity to provide responsive and accessible ways to create, observe, transform, archive, and re-create "folds" (Deleuze 2006; Deleuze and Hand 1988) of urban fabrics and cityness (Sassen 2005) today. We are indeed aware that battalions of technical designs with similar aims for engagement and "enhancement" of people's lives have been deployed but failed to meet their expectations and/or objectives. Various master planned communities and so-called smart cities across the world are some of the most obvious examples. We are also aware that it can be at times pernicious to simply roll out technologies simply because it is possible to do so.

Given this climate of complexity and urgency, we argue that designing for resilient futures must embody a deep understanding and respect for *people's needs and desires* and the historically formed underpinning of the *city as a place of diversity and contestations* giving rise to a wide range of expectations and hopes. Further, we argue that technologies as increasingly embedded in landscapes and people must therefore be *playful* as defined by Choi (2010): open, messy spaces capable of engendering engagements in new forms of re-creative interactions that lead to changes in sociocultural values and broader systems. Designing at the intersection of human-computer interaction (HCI) and human-food interaction would have flavors that are primarily related to conditions and actions associated with food, but it must at its core, mix into it these three values. Researching at this intersection then inevitably requires the mixing and blending of disciplinary frameworks and methods. It is time for a shift in how we work together toward a sustainable and just future and explore opportunities for a different kind of well-being that is achieved when people: feel empowered to be aware of the current systems and configurations of their environments; explore possibilities for change and voice their thoughts independently; and have accessible and profound means to collaborate to pursue realizing the possibilities.

Within this broad vision, our ambition for this book is modest. We see it as a small beginning to discussions about how we understand and design for human-computer interactions for sustainable human-food interactions. We are genuinely grateful for and delighted by all the valuable contributions our colleagues have kindly made to this book. We carefully hand-picked writings that address the three values proposed above but in particular, that of *people*. As the reader will see, the chapters are arranged in three themes—eat, cook, and grow—and the authors take various research methods and pose equally varied questions. However, there are many common subthematic and conceptual threads that weave the chapters together. The chapters will be introduced in the section forewords kindly contributed by our colleagues, whose unique perspectives and methodologies are deeply appreciated by us and wider communities of researchers and practitioners: Anne Galloway, whose creative research methods include fascinating speculative design work exploring future scenarios of New Zealand merino wool; Yvonne Rogers and Kenton O'Hara, whose *in the wild* research has produced numerous innovative playful systems; and Eli Blevis, whose pioneering research in sustainable interaction design and visual thinking has had a profound influence on a myriad of developments within the HCI field. Without further ado, we present to you this mix-max bibimbap of a book, with its distinct flavors of thoughts working together to create a unique experience mixing human-computer interactions and human-food interactions.

MIX-MAX BIBIMBAP

Makes a variable number of servings
When it comes to Korean cuisine, diversity is undoubtedly one of its main qualities. A standard meal consists of rice, soup and/or stew, and various side dishes known as banchan in Korean. Each individual seated at the table has his or her own bowl of rice and soup even though diners commonly share the meal, taking and eating portions of food directly from the others' bowls. Here, multiple dipping is not only permitted but necessary. The epitome of such meal sharing is bibimbap (mixed rice): rice and various ingredients mixed together in a bowl with chili paste. In many cases, there is no set recipe; whatever is available right there and then—including banchan and leftovers in the fridge or pantry—can function perfectly as "legitimate" ingredients. According to Uh-ryung Lee (2006), the former minister of culture and tourism in South Korea, this culture of mixing and sharing—what he calls the co-supportive "chopsticks culture"—sustains and intensifies the current paradigmatic shift from IT (information technology) to RT (relation technology), connecting socio-emotional dimensions with those of the technological.

Ingredients

Cooked rice, typically brown, black, or a multigrain mix, perhaps with beans or even quinoa—whatever you fancy—although sticky rice may not be the best choice

Vegetables (cooked or uncooked), tofu, meat (beef or chicken)—whatever and how much ever you fancy

Gochujang (Korean chili paste)

Sesame oil and seeds

Preparation

1. If you have a stone pot, heat it up on the stovetop with about a tsp. of sesame oil and a cup of cooked rice. When you start to hear the rice make a crackling or sizzling noise, it will be time to mix in everything else. If you don't have one of these pots, a big bowl will do; just place the rice in the bowl and drizzle on some sesame oil.

2. On top of the rice put a dollop of gochujang. Feel free to add other heat-inducing delights such as hot pepper powder/sauce/oil.

3. Then gently spoon the rest of the ingredients over the rice, distributing them in pleasing way. You may like to add cheese or sashimi.

4. Sprinkle sesame seeds, nuts, or any garnish that tastes good and makes you happy.

5. The best way, remember, is to make a huge batch in a big bowl and share it with your loved ones.

Notes

1. See http://senseable.mit.edu/trashtrack.

2. See http://www.foodmood.in.

References

Arnquist, Sarah. 2009. Research trove: Patients' online data. *New York Times*, August 24. http://www.nytimes.com/2009/08/25/health/25web.html?pagewanted=all&_r=0.

Ashraf, Quamrul, and Stelios Michalopoulos. 2010. *The climatic origins of the neolithic revolution: Theory and evidence*. Williamstown, MA: Williams College.

Bar-Yosef, Ofer, and Anna Belfer-Cohen. 1992. From foraging to farming in the Mediterranean Levant. In *Transitions to Agriculture in Prehistory*, ed. Anne Gebauer and T. Douglas Price, 1–10. Prehistory Press.

Bell, Genevieve, and Paul Dourish. 2007. Yesterday's tomorrows: Notes on ubiquitous computing's dominant vision. *Personal and Ubiquitous Computing* 11 (2):133–143.

Bowles, Samuel. 2011. Cultivation of cereals by the first farmers was not more productive than foraging. *Proceedings of the National Academy of Sciences of the United States of America* 108 (12):4760–4765.

Choi, Jaz Hee-jeong. 2010. *Playpolis: Transyouth and urban networking in Seoul*. Unpublished PhD, Queensland University of Technology, Brisbane.

Choi, Jaz Hee-jeong, and Adam Greenfield. 2009. To connect and flow in Seoul: Ubiquitous technologies, urban infrastructure and everyday life in the contemporary Korean city. In

Handbook of research on urban informatics: The practice and promise of the real-time city, ed. Marcus Foth, 21–36. Hershey, PA: IGI Global.

Deleuze, Gilles. 2006. *The fold: Leibniz and the Baroque*. London: Continuum.

Deleuze, Gilles, and Sean Hand. 1988. *Foucault*. Minneapolis: University of Minnesota Press.

FAO. 2001. Rome Declaration on World Food Security. *World Food Summit*. http://www.fao.org/docrep/003/w3613e/w3613e00.htm.

Greenfield, Adam. 2010. Frameworks for citizen responsiveness: Towards a read/write urbanism. *Urban Omnibus*. http://urbanomnibus.net/2010/07/frameworks-for-citizen-responsiveness-towards-a-readwrite-urbanism.

Huizinga, Johan. 1955. *Homo Ludens: A Study of the Play-Element in Culture*. Boston: Humanitas.

Kimball, Michael. 1998. Global change from local decisions: An archaeologist's perspective. *Human Ecology Review* 5 (2):19–24.

Ledford, Heidi. 2010. Garage biotech: Life hackers. *Nature* 467 (7316):650–652.

Lee, Richard. 1968. What hunters do for a living, or, how to make out on scarce resources. In *Man the hunter*, ed. R. Lee and I. DeVore, 30–48. Chicago: Aldine Publishing.

Lee, U.-R. 2006. Digilog manifest (Digilog suhnuhn). Seoul: Senggak eui Namoo.

Malone, Evan, and Hod Lipson. 2007. Fab@Home: The personal desktop fabricator kit. *Rapid Prototyping Journal* 13 (4):245–255.

Montanari, Massimo. 2006. *Food is culture*. New York: Columbia University Press.

Price, T. Douglas, and Ofer Bar-Yosef. 2011. The origins of agriculture: New data, new ideas: An introduction to supplement 4. *Current Anthropology* 52 (S4):S163–S174.

Sahlins, Marshall. 1972. *Stone Age economics*. Chicago: Aldine-Atherton.

Sassen, Saskia. 2005. Cityness in the urban age. *Urban Age Bulletin* 2 (Autumn).

Sassen, Saskia. 2011. Open source urbanism. *Domus*. http://www.domusweb.it/en/op-ed/2011/06/29/open-source-urbanism.html.

Sauer, Carl. 1952. *Agricultural origin and dispersals*. New York: American Geographical Society.

UNFPA. 2007. *State of world population 2007: Unleashing the potential of urban growth*. New York: United Nations Population Fund.

Wirth, Louis. 1938. Urbanism as a way of life. *American Journal of Sociology* 44 (1):1–24.

Wei, Jun, and Adrian David Cheok. 2012. Foodie: Play with your food promote interaction and fun with edible interface. *IEEE Transactions on Consumer Electronics* 58 (2):178–183.

Wolinsky, Howard. 2009. Kitchen biology. *EMBO Reports* 10 (7):683–685.

EAT

Anne Galloway

It is no exaggeration to claim that one of the central linchpins of culture is eating with others. Although traditions and technologies for sharing food are as old as humanity itself, since the late 1600s Latin-based cultures and languages have used the word "conviviality" to describe the pleasures of eating and drinking in good company. The importance of this experience can be found in the etymology of the word, which literally means "living together." Although cultural traditions and technologies have changed over time, eating with others remains one of our most cherished activities. The chapters in this section describe some of these changing traditions and technologies, but always remind us that where our food comes from and what we do with it is inextricably connected to who we are and how we live in the world—together.

In "A Relational Food Network: Strategy and Tools to Co-design a Local Foodshed," eating is situated at the intersection of community and environment. The design project presented in this chapter highlights contemporary interest in local food production and consumption as it seeks to create rich interactions between farmers and the city-dwellers who eat their food. In this case, digital technologies are used to support both distance collaborations and direct communications, and emphasis is placed on the importance of human-digital platforms where the "last mile" of any interaction is a "human mile."

In "Technologies of Nostalgia: Vegetarians and Vegans at Addis Ababa Café," eating involves the creation, and maintenance, of both offline and online communities. The ethnographic project described in this chapter details the multicultural communities that form around eating at a local Ethiopian restaurant. Of particular interest is how experiences of eating together in the café, and how these experiences are represented online, act as a means for people to reflect upon similarities, negotiate differences, actively (re)construct both personal and cultural identities, and bridge the space between "tradition" and "innovation."

In "What Are We Going to Eat Today? Food Recommendations Made Easy and Healthy," eating is something that people associate with improved health and physical well-being. This chapter describes a prototype support tool designed to help people choose healthier ingredients and meals. While successful in terms of promoting greater awareness of healthy eating habits, and with a majority of study participants preferring to use a digital application, the researchers found that social relationships and shared meals can remain significant barriers to changing personal eating behaviors.

In "Not Sharing Sushi: Exploring Social Presence and Connectedness at the Telematic Dinner Party," eating is interpreted as a shared activity that can use communication technologies to "make present" people who would otherwise be absent. The project described in this chapter is designed around the shared "structures and dynamics" of a dinner party, including sight, sound, and the ability to offer food to others. Notably, the research suggests that while networked technologies can adequately offer experiences that are "mirrored" from one location to another, they may fail to offer the sense of simultaneous and immediate experience that people often associate with sharing food.

In "Civic Intelligence and the Making of Sustainable Food Culture(s)," eating is positioned as the practice of sharing food within geographic communities. This chapter emphasizes local, sustainable agriculture and food cultures that can benefit from the use of communication technologies to "organize, communicate, and assemble in opposition" to global industrial agriculture. Notably, online networks are seen to be critical elements in supporting the kind of civic intelligence needed to reconfigure dominant modes of production and consumption offline.

Throughout all these stories of technology design and use, eating is identified, and indeed valued, as a fundamental act of sharing food with others in highly situated and embodied ways. A desire to create and support more sustainable and convivial choices drives all these projects, and the most successful uses of technology do not seek to replace existing eating experiences, but instead offer the opportunity to forge new, complementary relationships among people, places, and food. As we can see in this section, a thoughtful approach to eating and technology use allows us to reshape not only our understanding of who we are but also what we want our relationships with environments, economies, and politics to be.

SUMMER VEGETABLE AND CHORIZO STEW

Makes 6–8 servings
This is an easy and tasty one-pot meal that can be doubled or made in even larger batches. Omit the chorizo and substitute vegetable stock for vegetarian version.

Ingredients

2 yellow onions, diced

1 or 2 fennel bulbs, sliced

2 or 3 red peppers, chopped

4 or 5 zucchini, chopped

olive oil

Harissa (hot red pepper and spice paste used in North African cooking; Tunisian varieties are widely available in urban grocery stores)

4 to 6 potatoes, peeled and cubed

1 15 oz. can of chopped tomatoes

3 or 4 links of dried chorizo, chopped (optional)

32 oz. chicken (or vegetable) stock

Feta cheese, crumbled, for garnish

Preparation

1. Sauté onions and fennel in olive oil. Add red peppers and zucchini to pot, and continue to cook.
2. Once the vegetables have softened, add a generous amount of good quality harissa, the potatoes, and the tomatoes.
3. Cover with stock and bring to a boil. Reduce heat, add chorizo, and simmer for an hour.
4. Lightly mash the stew before serving, and top with feta cheese.

1 A Relational Food Network: Strategy and Tools to Co-design a Local Foodshed[1]

Joon Sang Baek, Anna Meroni, and Giulia Simeone

This chapter reports on an ongoing project aimed at starting up a network of local food services and sustainable production and consumption practices linking the city of Milan with its peri-urban area. "Feeding Milan: Energies for Change," as the inititive is called, aims to transform Milan into a local foodshed, a real-scale living experiment of a sustainable food system.

In the chapter we discuss the reasons why a relational and convivial strategy has been chosen to start this process of cultural and technical transformation. Then we describe what tools are being used to involve the different stakeholders in an extensive participatory design strategy, whose "last mile" of interaction is always a "human mile," even though supported by digital technologies. After offering an overview of the services currently under progress, we turn to the design and prototype of the digital platform and, in particular, of the vegetable foodbox program.

Designing for Territorial Development

Contemporary design—in its multiple articulations of strategic design, design for services, and product-service system design—has interwoven strong connections with the disciplines of territorial development and planning, especially from the perspective of sustainability (Manzini 2010b; Moy and Ryan 2011; Ryan 2009; Thackara 2005). These connections are largely due to the holistic approach that is theorized and actually put into practice when design considers "place" as a crucial component of an intervention (Meroni 2008; Zurlo 2010) and when design action touches the vital relationship of a community with its environment, its sense of belonging to a space, and the relationship between local people and resources. Taking various and nuanced perspectives, some schools of urban and regional planning around the world make these connections the focus of their investigation. Such schools,[2] many of which are recognized as scientific authorities in scholarly discussions about sustainable place development, (Meroni 2011), believe that valorizing local heritage (in terms of both environment and culture) is the only possible approach to producing lasting enrichment in a geographic area, since places are the result of the historical co-evolution of a human settlement (society) within its natural environment.

This leads us to use the term *territory* according to the Italian meaning of the word as it refers to the unity of a place and the way in which the people act and exist within it as they develop a sense of belonging and ownership. Sustainable development not only refers to the reproducibility of natural resources, but also to the way in which urban systems are established; to the coherence of production systems with local resources and entrepreneurship; to the development of capability and self-government by local communities (Magnaghi 2000). This multilayered definition of *territory* is the basis from which scholars and professionals in the fields of strategic and service design have recently taken up the debate about regional development (Meroni, Simeone, and Trapani 2009; Jégou 2010; Manzini 2010a). They look at the interrelations within a community and at the relations of the community with its *territory*.

Designing for sustainability means facing and promoting changes in ways of living (SPREAD 2011): we acknowledge that the great challenge of changing behavioral paradigms calls for action that touches on cultural values, making present changes and possible future behaviors manifest, and exemplifying systemic changes at the level of everyday experiences.

Changing behavioral paradigms requires collaborating with innovative social parties in social innovation processes; it implies involving the most proactive and creative resources in the design activity itself (Meroni 2007). Design for social innovation is the use of design thinking and knowledge to foster change in the behavioral sphere by supporting existent initiatives and helping new ones to begin. It is a mix of strategic and service design, where special emphasis is put on collaboration and participation, and whose goal is sustainability.

In this chapter we present a project that testifies to this contemporary way of designing. Although it makes no claim to exemplify the model, it effectively encapsulates a series of methodological and contextual conditions that enable us to reflect on this way of interpreting and undertaking design today.

Feeding Milan: Designing a Local Foodshed

Nutrire Milano: Energie per il cambiamento (Feeding Milan: Energies for Change, www .nutriremilano.it) is an ongoing applied research project, started in January 2010, which aims to restore the relationships between the city and its productive countryside by creating a network of "0 miles"[3] food services, so as to make the metropolitan region a sustainable foodshed. Funded by Fondazione Cariplo (a Milanese bank foundation) and by the local government (municipality and province), Feeding Milan is a combined venture of Slow Food, Politecnico di Milano, and Università delle Scienze Gastronomiche (University of Gastronomic Sciences). The project came to life in response to the upcoming 2015 International Expo, which will be hosted in Milan and will focus on sustainable food systems.

Partner expertise includes food culture and gastronomy, agriculture, service, and communication design. Slow Food is an international association founded in 1989 as well as a cultural movement that proposes a philosophy of pleasure and sustainability, and which advocates educational programs about taste discernment in the name of a food that was "good, fair, and clean" (Petrini 2007). At the Politecnico di Milano, work on the project is conducted in the Design department, the institution's design research and education center. The University of Gastronomic Sciences is an international research and educational center for a holistic relationship between agronomy and gastronomy.

Figure 1.1
The poster of the project. Image by Department of Design—Martino Cazzaniga and Francesca Piredda.

Context and Reasons for the Project

Feeding Milan is set in a rur-urban region (Donadieu 2005): a fringe area challenged by building speculation, where urban sprawl is blurring the boundaries between city and countryside and agriculture is suffering because it is no longer profitable. Such a lack of identity and vision is highly risky in terms of land exploitation and social cohesion. It calls for radical thinking and systemic change in the way we look at relations between peri- or rur-urban agriculture and the city.

Despite Milan's location in one of the biggest agricultural areas in Europe, the Agricultural Park South, the demand for fresh local produce in the city exceeds what is currently available. In fact, the park, a blend of living/productive settlements and fields, covers an area of 47,000 hectares and is characterized by intensive industrial farming; a very small percentage is dedicated to diversified agriculture and eco-compatibility systems. Besides its merely productive resources, the Milanese region has an important historical heritage, rural villages, water resources, and a rich cultural mix. All these resources could be turned into assets to rethink the identity and the nature of the area.

"Feeding the planet. Energy for life" is the motto for the International Expo 2015, highlighting the importance of a sustainable land use for food provision in the near future. The project Feeding Milan had, since the beginning, the ambition to start a real-scale experiment, by converting the principles of the Expo into a "regional monument" to sustainable food; in other words, to create a living good practice rather than just a showcase. To do this, the project is boosting agriculture as a presidium of the land's quality, revitalizing local networks, encouraging good practices, and optimizing the use of resources.

Objectives

The project moves from the assumption that only by creating an efficient and effective agri-food chain that is sustainable and self-sustainable, will it be possible for peri-urban agriculture to flourish, feeding the city and offering city dwellers a multiplicity of nature-related activities. Feeding Milan, therefore, operates so as to foster the development of a local and multifunctional agricultural system by offering an integrated network of services to both farmers and citizens. In detail, the goals of the project are the following:

• To promote a shift from industrial agriculture based on monoculture and intensive farming to organic multifunctional farming. This is by supporting existing good practices and by developing new, specifically conceived, projects.
• To implement a set of services to improve the quality of life in the countryside, and to facilitate the links between farms and the city.

• To activate unexploited resources that the territory may offer.
• To promote a new culture of food and agriculture and to raise awareness among citizens of the importance of peri-urban agriculture for the quality of life in Milan.

Scenario and Services: Planning by Projects

This project draws competencies and tools from the disciplines of strategic design (for issues that concern stakeholder involvement and quality of the processes) and service design (for issues that concern conceiving, structuring, and modeling the service activities). It involves an approach best defined as planning by projects, or acupunctural planning (Manzini 2010a; Jégou 2010; Ryan et al. 2010): a form of territorial planning based on pinpointed interventions on a micro scale, where single projects are grafted into an interconnected network. Stakeholders already proven to be proactive in the region are engaged to activate the new projects; initiatives are then connected to create synergies (Meroni, Simeone, and Trapani 2011), configuring a system that shares resources and brings about economies of scale and purpose.

At the present stage of the work, several local projects have been activated. Some of them are running as almost-consolidated experiments, others are in the prototyping phase, and others are about to be implemented. This list provides an overview:

• *Il Mercato della Terra di Milano* (The Milano Earth Market): a monthly farmers' market, the first to operate in Milan public space, established in December 2009. It is a rich, complex social environment besides being a point of sale.
• *La Cassetta del Contadino* (The Farmer's Foodbox): a weekly delivery of fresh local food with a logistic system based on pick-up points in various neighborhoods. The foodbox is currently in the first stage of operation.
• The Super-Coop: a concept of a collaborative supermarket totally managed by customers. It is a contemporary model of the consumers' cooperative, where the members collaborate to run the service so to keep prices as low as possible.
• Yes Weekend: the park's tourist agency has created the first in a planned series of do-it-yourself services for local tourism in Agricultural Park South.
• Bike Pass—Bike Sharing in the Agricultural Park: a system of routes and bike-sharing stations in the farmhouses of the park, connected to the Milan public transport network. Partially tested, preliminary routes are up and running.
• The Pick Your Own: a network of farms where you can pick and buy fruit. The service is now at the conception stage and will be possibly organized as a community-supported system.
• The Local Bread Chain: bread entirely produced in the region, from the wheat to the loaf. The first crop was harvested with pilot farmers in summer 2011 and the first bread was sold in the Earth Market in September 2011.

Figure 1.2
A moodboard of the various services currently activated within "Feeding Milan." Pictures by
Department of Design's team.

The Project Philosophy: Making Things Happen through Conviviality

Local projects are a way of starting bigger behavioral changes. Designed to make things
happen quickly, to be effective, and to gain the trust of the community, they are conceived so as to spark a broader process of transformation that can be supported by
collaboration and based on emulation (Drayton and Budinich 2010). Changing behavioral and productive paradigms is labored and risky; it does not happen if the stakeholders do not find convenience, commitment, and trust, which together spur a
community to try out new possibilities. "Feeding Milan" is experimenting with two
key principles to lead this transformation.

The first one is conviviality. Conviviality is not only about eating together, but it
is actually about creating pleasurable and collaborative relationships in every activity
of life. In Latin cultures, food means conviviality, pleasure, taking care of others, and
being loved. Michael Pollan (2008) refers to conviviality as an expedient to moderate
consumption on one side and, above all, to benefit from the social approach that food
triggers though rituals, traditions, pleasure, and so forth. In this sense, the Slow Food
movement is not so much about cooking and eating more slowly as it is about con-

necting people and regaining meaning for the rituals related to food in everyday life. Food is the most powerful and "natural" tool for conviviality, as by "conviviality" we mean a condition of autonomous, creative, cheerful interaction between people. Conviviality in the Feeding Milan project is about building a network of trust and sympathy among the producers themselves and with the consumers. It encourages the "last mile" of an interaction to be a "human mile" during which relationships occur. This also implies that designers be part of, and create, the human links necessary to operate within such interactions. Conviviality is thus a way to challenge the current industrial retail system by making people experience the pleasure of indulging in relationships around food and using the gentle power of collaboration and feeling good together. The aim here is for conviviality to be one of the main reasons why producers and consumers change behavior: the project, in fact, strives to combine a high value in relationship with accessibility.

The second principle is the effectiveness of prototyping, or using design action to experiment with new ways of doing things. Feeding Milan started as a project aimed at making things happen: in design terms, this implies moving fast from the idea to the implementation through service prototyping and concrete actions. Experimenting with new ideas as soon as possible with the community is actually how the project works. To make this possible, a specific "tool" is used: the Ideas Sharing Stall, a booth in the farmers' market where new services are presented and discussed with the visitors by interacting and reacting on draft ideas elaborated by the design team.

Conviviality in Place: The Milan Earth Market and Its Ideas Sharing Stall
A brief introduction to Milan's Earth Market (a Slow Food farmers' market) will explain the convivial mood of the project. According to the principles of Slow Food and its already mentioned philosophy of "good, fair, and clean," the Earth Market is not a simple farmers' market: it is a place where people can hang around enjoying their Saturday morning. It offers a rich and multifaceted experience that people simply like: besides the food shopping and the face-to-face meeting with the producers, it proposes tasting sessions, workshops, eating occasions, and socializing spaces (a number of "conviviality tables" are provided for free use and are always populated by people eating food they buy in the market), as well as music and presentations. As a consequence, visitors enjoy staying longer and can indulge in doing things with care and *slowness*. Their food shopping becomes a meaningful activity rather than a duty "to get over and done with as fast as possible," as it is in ordinary supermarkets. The market aims to be a platform where producers and consumers regain the value of knowing food and speaking about food, which is a profound Italian peculiarity, and a venue where they can spend time occupied with food rituals and their significance. The Ideas Sharing Stall is an element that opens up new conversations: people manifestly enjoy interacting at this stall, participating and being part of a collective decision-making

process. Designers support and guide these conversations toward the birth of new services.

From a technical perspective, the configuration of this market responds to a concept of *multifunctionality*: the principle of combining several functions in the same activity lies at the root of the entire project, as it increases the self-sustainability of the services by integrating different sources of income, and enhances the quality of the experience for both producers and users (Forlizzi and Battarbee 2004; Battarbee and Koskinen 2005). It helps authentic, positive co-experiences to take place, giving participants the possibility of meeting up. The conviviality tables are one of the service touchpoints set up for this purpose: the simple fact of being there invites people to gather. They facilitate interaction and help to develop the human, collaborative platform of Feeding Milan.

Feeding Milan: The Digital Platform

As explained by the project philosophy and its particular context, Feeding Milan is a project based on interpersonal, collaborative relations and human contact. To relieve the system from some organizational burden and to make it possible to concentrate on the quality of the offerings and experience, a digital platform facilitates the interactions among the stakeholders.

Ever since the beginning of the project, it was clear that the smooth operation of the collaborative system at the basis of both the front and back stage of the services would benefit from a digital platform. It was equally clear that this platform would be superimposed on a positive, relational (human-to-human) network. Consequently, a website was developed with aims to create a sense of community among the project stakeholders and to empower them to fulfill their social and economic needs by providing the necessary tools. It was conceived with two roles: (1) to provide the local producers and consumers with a support platform for the creation and reinforcement of a social network that underpins the operation of services that, unlike commercial ones, are co-produced by producers and/or consumers (and hence is hereafter called collaborative service); and (2) to provide them with technical support to enhance communication and collaboration with digital tools.

The first role was based on the idea that Internet communication technologies (ICTs) create and strengthen social networks, mainly weak ties (Wellman, Quan-Haase, Witte, and Hampton 2001, Kavanaugh 1999). Weak ties, according to Mark Granovetter (1973), play a crucial role in the diffusion of innovations by connecting groups that would otherwise remain isolated. Likewise, in social innovation, local communities with creative solutions can be connected through weak ties, thereby creating a network of locally based sustainable solutions. According to Manzini (2010b), qualities such as *small*, *local*, *open*, and *connected* characterize a sustainable society. Therefore, if designed successfully, ICTs can facilitate a transformation toward a sustainable society by providing the means to build the social and technical infrastructure on which creative ideas are shared and collaborations initiated.

The second role relates to user-driven innovations facilitated by the democratization of ICTs. ICTs have been increasingly adopted in social innovations, including grassroots initiatives, to empower individuals to generate solutions for their own needs. Various digital tools for efficient communication and collaboration improve the functionality of their solutions. The Agricultural Park South of Milan is no exception. Milan and its surroundings have a relatively high penetration of the Internet compared to the national average, thus satisfying a basic condition for the proliferation of ICT-enhanced collaborative services. In fact, many producers and consumers are already using technologies to create communities and to promote local food consumption. The platform aims to make technology more user-friendly by providing services that motivate the producers to participate and by increasing consumers' access to local products. It aims to support both the functional and social aspects of collaboration in the network.

Platform Characteristics for Collaborative Services

Prior to developing a digital platform for the project, case studies were conducted to explore how individuals and communities use ICTs in their collaborative services. In these studies, we qualitatively analyzed over thirty websites and mobile applications in terms of the role of ICTs in supporting the operation of the services and reinforcing users' social networks. We discovered that, despite the diverse contexts of the cases, a common structural system existed. It consisted of four layers: a platform base; an enabling solution; a collaborative service; and an event (Baek, Manzini, and Rizzo 2010).

A platform base is a system-enabling solution that provides a set of tools and user information necessary for collaboration. It is what Quentin Jones (1997) calls a cyber region/locale and contains many virtual communities. An example of a platform is Meetup.com, a social networking service that facilitates face-to-face gatherings. On the Meetup platform, there are thousands of enabling solutions for collaborative communities, such as Team Fighting Diabetes (in California), one of many such communities formed by diabetics and their families.

An enabling solution refers to a set of products, services, and organizational tools that empower individuals or communities to achieve a result using the best of their skills and abilities. In this context, its scope is reduced to digital artifacts. For example, on Meetup.com, a diabetic can easily search for and join a local group of people who share the same illness.

An event is the manifestation of a collaborative service; that is, the physical evidence of a collaborative experience (Titz 2001). For example, the Team Fighting

Diabetes meetup group organized a hiking day for its members to socialize and exercise together.

Digital Platform Design

This common structural system suggests that a reference model could be proposed for conceptualizing a digital platform for collaborative service. Such a model would include the four layers and the configuration of digital tools to facilitate the production of collaborative services. The platform itself would account for the two bottom layers—a platform base and an enabling solution module—and the upper two layers would be a collaborative service and an event.

A platform base includes digital tools and a user profile database that serve as building blocks in an enabling solution module. The digital tools are divided into communication and noncommunication tools: communication tools include social media, multimedia, and online broadcasting; and noncommunication tools include maps, global positioning systems (GPS), an e-commerce system, and search engines. The user profile database consists of data relevant to the services supported by the platform, such as the producers' location, type of product, skills, tools, knowledge, and interests.

An enabling solution module is equipped with tools for organizing and maintaining a collaborative service. The platform uses the modular design so that all enabling solutions on a platform base share the same tool repository and the user profile database, and new modules can be added as new services are launched (Voss and Mikkola 2007). As the Feeding Milan project proceeds, the tool repository will expand, and new users will join in, thus achieving economy of scale in the production of the services.

For example, a foodbox delivery service (The Farmer's Foodbox) was designed to supplement the farmers' market. It targets consumers who are willing to purchase local

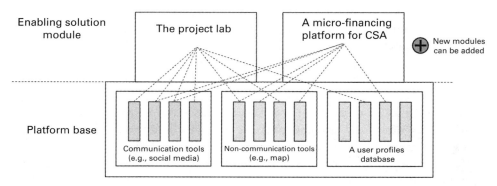

Figure 1.3
The structural system of a digital platform for collaborative service.

food but have difficulty accessing the market and those who are not satisfied with the existing foodbox schemes. An ideal enabling solution for foodbox delivery consists of an e-commerce system; a blog that allows producers and consumers to communicate; an online supply management system; a map to show the location of producers and proximity points; and a search engine to find producers, products, price range, and location. The following section discusses the development of the foodbox delivery service in more detail.

The Farmer's Foodbox: A Design Experiment

The Farmer's Foodbox (*La Cassetta del Contadino*) is a weekly delivery service of local, fresh vegetables. It exemplifies how the integration of the human network with the digital platform can be managed through a collaborative principle. The service back-stage is organized around country hubs that collect the produce from farms all around their own area and pack them into boxes. A carrier brings the boxes to a number of proximity points in the city—corner shops or offices, for example, which are conveniently located for the hubs—and distribute the boxes to the final consumer. The digital platform manages the interactions.

Service Development

Designing this service took several steps. After a first phase of problem setting and context analysis, a concept was outlined. It was first developed in a co-design activity at the Ideas Sharing Stall of the farmers' market. According to the result of this evaluation, the design team developed the event to be tested in a prototyping phase. The aim of the prototype was to assess person-to-person interactions in the network of people and the management of orders, information, and economic transactions in the digital sphere of the platform (Simeone and Corubolo 2011).

The service prototype was shaped in view of the opportunities and constraints of an experiment, which means that some of the initial features had to be put aside in favor of newly emerging conditions. Designers had to go through the following steps in order to develop the project:

• *Context and user understanding.* The approach to designing the service was based on proposing ideas: rough concept prototypes were discussed at the Ideas Sharing Stall to better understand people's needs and expectations. This activity helped the design team to orientate the next development phase. Interviews with people already using other similar services were conducted. This benchmark helped to place the service offering within the given market.

• *Co-design sessions.* The whole Feeding Milan team was involved in a series of design sessions aimed at setting up service priorities and conceiving an appropriate

organizational model. Producers were involved to define the foodbox quality standards. The brand image was developed according to the Feeding Milan identity.
• *Prototyping.* A first pilot experiment was set up in June/July 2011. Around fifty pilot
users were engaged, and three country hubs and five proximity points were activated
to deliver around 650 kilograms of produce. Farmers of the country hubs were trained
to sort and package the boxes, while the people at the pick-up points were helped
with managing the logistics. The run time was monitored through the digital platform
to collect continuous feedback from the users and make small adjustments on the run.
Several operations that were supposed to be automatically managed by the digital
platform were in reality "crafted" by simulating procedures.

At the end of the first pilot test, some considerations on the service prototyping
phase and the service itself emerged:

• A "zonal" service, connecting, on one hand, a country hub and a number of pick-up
points which are closeby and, alternately, users living in a same neighborhood, can
work well. The proximity of the pick-up point with users facilitates the creation of
personal relations among them as well as the beneficial links to the keeper of the point
(i.e., new business or social interactions).
• The number of stakeholders involved in the pilot must be relatively small in order
to be able to control and check the whole process and intervene in the case of critical
problems. Too many inexpert actors working together can result in mistakes challenging the experiment.
• The training of the service providers is an essential step, not to be neglected, in
order to guarantee a standard quality to the users; in fact, all stakeholders need to
understand the advantages, the constraints, and the rules of being part of such a
system.

While a number of lessons on the specific nature of a collaborative service were
learned, some clear insights into the role of a digital platform emerged. In particular,
we can mention the utility of using technology for the management of the quantita-
tive and functional requirements of the service, and for the creation of direct communication channels between users and farmers. Whereas these functions are
conveniently managed via digital interaction, it is crucial that the last touchpoint of
the user and the system was human. When the user picks up the foodboxes from
someone at a designated point, this contact person responds to user feedback about
the quality of the whole service and establishes a moment of trust, dialogue, and
spontaneous sociability. Even when the user happens to express disappointment about
some aspect of the service, this moment has great value as a way to stimulate relations
at the local level of the neighborhood. This underlines, also, the importance of creating a sense of service ownership among all the stakeholders and the need to set up a
collaborative environment that enables them to learn from their own experiences. The

digital platform has proven to be effective in facilitating this collaborative environ- ment, thus helping the various stakeholders exchange feedback about the service and improve its quality.

Conclusions: A Relational Network

Feeding Milan is geared to offer accessible solutions to a wide range of citizens, and, as such, requires a relational network—an interactive and collaborative effort—for success. The following sections assess and summarize its platform design.

A Human-Digital Platform for a Relational Network

As we discuss above, the accomplishment of this mission requires setting up a col- laborative platform and involving numerous stakeholders to work together. The design experiments so far conducted teach a tough lesson about how critical some elements can be. These include: reaching and maintaining a standard level; assuring continuity; managing the limits of production and processing capacity; and receiving clear and direct feedback from the consumers and the producers.

The project also illustrates the importance of a process based on fast prototyping: collaborative services, whether supported by digital platforms or not, need reciprocal understanding and trust, which come from a growing capacity to do things together. By engaging together in an activity, the community can grow in competence and reciprocal respect, despite initial differences and inexperience. This way, weak ties grow stronger and create the necessary social network to make things go well. Involving those proven to be proactive in social innovation in new, experimental activities has been effective in creating a working platform, spreading and scaling up the businesses by starting positive processes of emulation. They help to build trust around the initia- tives and push the community to become self-entrepreneurial and braver.

Feeding Milan is a platform for action that mixes task-based and fun-based inter- actions (Hassenzahl 2003; Forlizzi and Battarbee 2004; Battarbee and Koskinen 2005) so that initiatives are always driven both by need and enjoyment. In other words, people are largely moved by the pleasure of doing things for "their own sake" and the discovery of themselves not just as consumers, but as individuals determining their own lives (Inghilleri 2003; Csíkszentmihályi 1990). In open and collaborative service models these two behaviors are often mixed, so that designing the tools and the rules of the interaction is a major challenge. The designer's job here approaches that of a coach: to train people to perform a function, to interact with others, and to create productive collaboration.

The digital platform can facilitate these tasks, but nevertheless it has been observed that to positively develop new practices in the community, the members have to meet and talk face-to-face. The Ideas Sharing Stall at the farmers' market is a crucial contact point for the whole community and is often the kickoff for further interactions that take place in either the digital or physical world.

Design Role: Is the Designer an Activist?

Through the project Feeding Milan a new and different nature of design becomes fully apparent. The role of a designer, as suggested above, has evolved toward that of a community coach: someone able to discuss and share ideas, using professional tools to make things happen and enable people to take action.

By choosing the concept of conviviality to inform the practice of designing, the project gives emphasis to the importance of the designer "being there," spending time within the community and participating in person in the process of change. In doing this, a designer serves as a social activist supporting social innovation processes within the community and becoming part of it, for as long as it takes, until initiatives become self-sufficient and the community "competent" (Meroni and Sangiorgi 2011). Here the designer's role as an observer of (and a participant in) people's lives includes motivating actions, mobilizing stakeholders, and creating "spaces of contest" that reveal and challenge existing configurations and conditions of society (Markussen 2011).

Finally, as a mix of human and digital interaction, the Feeding Milan platform can be seen as a tool for developing society's capacity to effect change, through which citizens, farmers, and academics work together in a culture of continuous transformation and self-criticism. Echoing Ivan Illich (1973), the project aims to increase the malleable quality that makes possible reciprocal understanding and respect, and to increase the user's autonomy; the capacity to promote autonomy can be considered in itself a fundamental characteristic of convivial tools.

Acknowledgments

The Feeding Milano project is the result of the efforts of a big team whose main participants are: Anna Meroni, Giulia Simeone, Daria Cantù, Marta Corubolo, Francesca Piredda, Daniela Selloni (Politecnico di Milano), Alberto Arossa, Paolo Bolzacchini, Alessandro Cecchini (Slow Food Italia), Paola Migliorini, and Elio Nasuelli (Università Scienze Gastronomiche).

PACIARELLA (VILLAGE CAKE)

Makes at least 8–10 servings

The Lombardy region is well known for producing the two main ingredients of this cake, milk and bread. Village cake originated as a "poor man's cake" to use up leftover bread by soaking it in fresh milk. No single "right" recipe evolved; every family would add other ingredients (such as cocoa, raisins, and pine nuts), according to what was at hand.

Making paciarella is a long and slow process that requires patience to wait while it rests. Traditionally, people used to bake the cakes with the collaboration of the village bakery, as they couldn't afford a family oven. Folks enjoy this cake every first Sunday of October, to celebrate the autumn and the village as well. Having said that there is no single "right" recipe, we offer one from our family.

Basic Ingredients

14 oz. stale bread

6 cups fresh milk

5 oz. raisins

7 oz. dark cocoa

7 oz. crumbled amaretti (almond biscuits)

Anise and vanilla (to taste)

¾ oz. pine nuts (for topping)

Optional Ingredients

The following ingredients have become our family favorites. Please follow our suggestion or add personal favorites to make your own cake special.

1¾ oz. candied citron

9 oz. shortbread, crumbled

7 oz. sponge cake or ladyfingers, in pieces

Preparation

1. Place small cubes or torn pieces of stale bread in a large bowl. Pour in hot milk and mash it with a wooden spoon. Let it rest at least 2 hours. In the meantime, let the raisins soak in water till swollen.

2. After 2 hours, in the same bowl of bread and milk, add raisins, the biscuit and/or cake crumbles, the citron, cocoa, vanilla, and anise (to taste).

3. Stir everything well with a wooden spoon and let it rest for at least four hours (but preferably overnight). If the mixture is too stiff, add some more milk until it looks creamy.

4. Grease a deep 9-inch cake pan with butter so the cake won't stick. Pour the mixture in the pan and sprinkle pine nuts on the top, as decoration.

5. Bake for 10 minutes at 400°F (200°C) and then turn down the temperature to 200°F (100°C) for 2.5 hours. This will prevent the cake from becoming too hard on the top. To check for doneness, pierce the cake with the tip of a sharp knife or a toothpick: if it comes out dry, the cake is ready. Let it cool before cutting, and then enjoy.

Notes

1. This paper reports on collective work and is the result of collective reflection. Nevertheless, for the purpose of this publication, Joon Sang Baek co-wrote the section "The Digital Platform"; Anna Meroni wrote the sections "Designing for Territorial Development" and "Conclusions," and co-wrote "Project Philosophy" with Guilia Simeone; and Giulia Simeone wrote "Feeding Milan" and "The Farmer's Foodbox."

2. These include the Scuola Territorialista in Italy, the New Urbanism movement in North America, and INTBAU, the International Network for Traditional Building, Architecture & Urbanism in the UK. Scuola Territorialista (Magnaghi, 2000) is a branch of Urban Planning organized at the beginning of the 1990s by a group of urban planning and sociology scholars who decided to coordinate their work under common principles.

3. The expression "0 miles" food production and consumption is a rhetorical hyperbole used in several international contexts ("Km-0" in Italy, "Zero-mile food" in English-speaking countries) to define as the shortest possible geographical and market food chains. For this particular project it refers to an area of 40 kilometers around the city of Milan.

References

Baek, Joon Sang, Ezio Manzini, and Francesca Rizzo. 2010. Sustainable collaborative services on the digital platform: Definition and application. Paper presented at the Design Research Society International Conference, Montreal, Canada, July 7–9.

Battarbee, Katja. 2004. *Co-experience: Understanding user experiences in social interaction.* Helsinki: UIAH. https://www.taik.fi/kirjakauppa/images/2be572c773f32c5b5450d0b313a02c65.pdf.

Battarbee, Katja, and Ilpo Koskinen. 2005. Co-experience: User experience as interaction. *CoDesign* 1 (March):5–18.

Bjorgvinsson, Erling, Pelle Ehn, and Per-Anders Hillgren. 2010. "Participatory design and democratizing innovation." In *Proceedings of the 11th Biennial Participatory Design Conference*, 41–50. New York: ACM.

Brown, Tim. 2009. *Change by design*. New York: HarperCollins.

Cantù, Daria, and Giulia Simeone. 2010. Feeding Milan. Energies for change. A framework project for sustainable regional development based on food de-mediation and multifunctionality as Design strategies. In *Proceedings of the Cumulus Shanghai conference 2010: Young Creators for better city and better life*, ed. Yongqi Lou and Xiaocun Zhu, 457–463. Helsinki: Aalto University School of Art and Design.

Csíkszentmihályi, Mihaly. 1990. *Flow: The psychology of optimal experience*. New York: Harper and Row.

Donadieu, Pierre. 2005. *Campaignes urbaines*. Arles, France: Actes Sud, Ecole nationale superieure du paysage.

Drayton, Bill, and Valeria Budinich. 2010. A new alliance for global change. *Harvard Business Review* 88 (September):57–64.

Forlizzi, Jody, and Katja Battarbee. 2004. Understanding experience in interactive systems. In *DIS '04 Proceedings of the 5th Conference on Designing Interactive Systems: Processes, Practices, Methods, and Techniques*, 261–268. New York: ASM.

Granovetter, Mark. 1973. The strength of weak ties. *American Journal of Sociology* 78 (May): 1360–1380.

Hassenzahl, Marc. 2003. The thing and I: Understanding the relationship between user and product. In *Funology: From usability to enjoyment*, ed. Mark A. Blythe, Andrew F. Monk, Kees Overbeeke, and Peter C. Wright, 31–42. Dordrecht: Kluwer Academic Publishers.

Illich, Ivan. 1973. *Tools for conviviality*. New York: Harper and Row.

Inghilleri, Paolo. 2003. *La « buona vita ». Per l'uso creativo degli oggetti nella società dell'abbondanza*. Milan, Italy: Guerini e Associati.

Jégou, François, and Stéphane Vincent. 2010. Residencies in public institutions supporting local transition to sustainable ways of living. In *Human cities: Celebrating public spaces*, eds. Barbara Goličnik Marušić, Matej Nikšič, and Lise Coirier, 31–35. Brussels: Stichting Kunstboek.

Jégou, François. 2010. Social innovations and regional acupuncture towards sustainability. Paper presented at the International Conference on Green Design, Zhuzhou, Hunan, China, May 21–22.

Jones, Quentin. 1997. Virtual-communities, virtual settlements & cyber-archaeology: A theoretical outline. *Journal of Computer-Mediated Communication* 3 (March).

Kavanaugh, Andrea. 1999. The impact of computer networking on community: A social network analysis approach. Paper presented at the Telecommunications Policy Research Conference, Alexandria, VA, USA, Sept 26–28.

Magnaghi, Alberto. 2000. *Il progetto locale*. Torino: Bollati Boringhieri.

Manzini, Ezio. 2009. *Casi e tipologie di incontro. Notes for the Service Design Studio*. Milan: Politecnico di Milano.

Manzini, Ezio. 2010a. The social construction of public space. In *Human Cities. Celebrating public spaces*, eds. Barbara Goličnik Marušić, Matej Nikšič, and Lise Coirier, 12–15. Brusselles: Stichting Kunstboek.

Manzini, Ezio. 2010b. Small, local, open and connected: Design research topics in the age of networks and sustainability. *Journal of Design Strategies* 4 (1):8–11.

Markussen, Thomas. 2011. The disruptive aesthetics of design activism: Enacting design between art and politics. In *Proceedings of Helsinki Making Design Matter: Nordes '11: the 4th Nordic Design Research Conference*, 102–110. Helsinki: Aalto University.

Meroni, Anna, ed. 2007. *Creative communities: People inventing sustainable ways of living*. Milano: Edizioni Polidesign.

Meroni, Anna. 2008. Strategic design: where are we now? Reflection around the foundations of a recent discipline. *Strategic Design Research Journal* 1 (January):31–38.

Meroni, Anna. 2011. Design for services and place development: Interactions and relations as ways of thinking about places: The case of periurban areas. In *Proceedings of Cumulus Shanghai Conference 2010: Young Creators for Better City and Better Life*, eds. Yongqi Lou and Xiaocun Zhu, 234–241. Helsinki: Aalto University School of Art and Design.

Meroni, Anna, Giulia Simeone, and Paola Trapani. 2009. Servizi per le reti agroalimentari: Il Design dei Servizi come contributo alla progettazione delle aree agricole periurbane. In *Produrre e scambiare valore territoriale: Dalla città diffusa allo scenario di forma urbis et agri*, ed. Giorgio Ferraresi, 161–200 Firenze: Alinea Editrice.

Meroni, Anna, Giulia Simeone, and Paola Trapani. 2011. Supporting social innovation in food networks. In *Design for services*, ed. Anna Meroni and Daniela Sangiorgi, 190–200. Farnham, UK: Gower Publishing Limited.

Meroni, Anna, and Daniela Sangiorgi. 2011. *Design for services*. Farnham, UK: Gower Publishing Limited.

Moy, Dianne, and Chris Ryan. 2011. Using scenarios to explore system change: VEIL Local Food Depot. In *Design for services*, ed. Anna Meroni and Daniela Sangiorgi, 161–171. Farnham, UK: Gower Publishing Limited.

Petrini, Calo. 2007. *Slow food nation. Why our food should be good, clean and fair*. Milano: Rizzoli.

Pollan, Michael. 2008. *In defense of food. An eater's manifesto. New York City*. The Penguin Press.

Ryan, Chris. 2009. Climate change and ecodesign, part II: Exploring distributed systems. *Journal of Industrial Ecology* 13 (3):350–353.

Ryan, Chris, Dianne Moy, Kate Archdeacon, and Michael Trudgeon. 2010. *Visions Broadmeadows 2032: Eco-Acupuncture. Enabling localised design interventions*. Melbourne: Victorian Eco Innovation Lab.

Simeone, Giulia, and Marta Corubolo. 2011. Co-design tools in "place" development projects. An ongoing research case. In *Proceedings of DPPI—Designing Pleasurable Products and Interfaces*, Milan: PoliScript—ACM In-Cooperation.

SPREAD. 2011. *Sustainable lifestyles. Today's facts & tomorrow trends*. Sustainable Lifestyles Baseline Report of the project SPREAD Sustainable Lifestyles 2050. http://www.sustainable-lifestyles.eu/fileadmin/images/content/D1.1_Baseline_Report.pdf.

Thackara, John. 2005. *In the bubble: Designing in a complex world*. London: MIT Press.

Titz, Karl. 2001. The impact of people, process, and physical evidence on tourism, hospitality, and leisure service quality. In *Service quality management in hospitality, tourism, and leisure*, ed. Jay Kandampully, Connie Mok, and Beverley A. Sparks, 67–83. New York: Haworth Hospitality Press.

Voss, Chris, and Juliana H. Mikkola. 2007. Services science: The opportunity to re-think what we know about service design. Paper presented at the Services Science meeting, Cambridge. June 14–15.

Wellman, Barry, Anabel Quan-Haase, James Witte, and Keith Hampton. 2001. Does the Internet increase, decrease, or supplement social capital: Social networks, participation, and community commitment. *American Behavioral Scientist* 45 (436):436–455.

Zurlo, Francesco. 2010. Design Strategico. In *Gli spazi e le arti. Volume IV: XXI Secolo*, 89–98. Roma: Enciclopedia Treccani.

2 Technologies of Nostalgia: Vegetarians and Vegans at Addis Ababa Café

Kit MacFarlane and Jean Duruz

"People say, 'This spice come from Ethiopia?' No, from here. From Australia."

"Eat my food the right way!"
—Yenenesh, founder of the Addis Ababa Café

Increasingly, online spaces are subject to criticism for simply affirming, rather than reworking, offline divisions and social structures. These spaces have been labeled as reactionary—"organized by race" and defined by "self-segregation" (Boyd 2010). The suggestion is that, in racial composition, the virtual world is fundamentally white, and simply restages offline cultural hegemony. Nevertheless, there remains the opportunity for such spaces to constitute interventions, even in the most modest ways, in mainstream social structures and cultures—to become sites for building alternative and transformative communities. In our chapter we explore this potential. If the online realm does indeed reflect the underlying tensions and concerns of everyday life, then this is also a realm that contains debates and challenges associated with the task of living "together-in-difference," a phrase coined by Ien Ang (2001, 200–201) to conceptualize the ambivalences and contradictions attached to a project of juggling our sameness with, and difference from, other people.

Addis Ababa Café, a small "ethnic" Ethiopian restaurant in Thebarton (a suburb of Adelaide, South Australia), provides us with a rich and resonant site for exploring interactions of sameness and difference, both within the café and in its connection to the local community.[1] This chapter draws on approximately three years of visiting, eating, observing, and conducting a series of ongoing interviews with staff, family, and customers; it will take you to a small African "hut" on a busy Adelaide highway (figure 2.1), and reveal a successful and multilevel engagement by an Ethiopian restaurant with its local community (we use "local" in its most literal sense, referring to the various and diverse cultures that have contact with the café and its surrounding environment). Additionally, the restaurant displays productive approaches to niche targeting of consumers via the online community of Facebook, making it an ideal site to

Figure 2.1
Welcome to Addis Ababa Café, a small Ethiopian restaurant in a suburb of Adelaide, Australia.

explore this juggling of sameness and difference through online and offline interactions. This ongoing engagement through the online realm not only carries basic marketing ambitions; it also, in the sense of Benedict Anderson's concept of "imagined communities" (1983), draws on cultural imaginings of an extended community around an Ethiopian presence, in both a physical and a digital environment.

Addis Ababa Café's success in creating an extended community with vegetarian and vegan consumers, propelled by its online interactions, has altered its initial approaches to its own consumer and community identity and engagement. This has provided a local Ethiopian identity vested both in "nostalgic place-making" (Hage 1997), on one hand, and new opportunities to build shifting and less usual political alliances, on the other. This presence within the online social networking space has extended the act of migrant "home building" through evocation of "nostalgic feelings" (Hage 1997, 105 and 108), while simultaneously allowing for accidental exposures that shape and reshape this home building. These exposures have allowed this seemingly traditional Ethiopian presence with a strong self-awareness of its own past and present identity to take advantage of opportunities to expand its role in community place making and the construction of its own identity. Online presence has not only expanded business connections, but allowed for the extended engagement with a vegetarian and vegan community, beyond mere marketing opportunities and without compromising a fundamental ethos and sense of cultural identity. Cultural change and transformation is thus not only reciprocal but also takes place through a careful interrelation of both the physical and virtual community spaces, transforming both and being transformed by them. In this chapter we intend to tease out this process.

"Making It Pretty": Physical, Digital, and Cultural Transformations

"Eat my food the right way!" Zed laughs, lovingly imitating his mother Yenenesh.

Zed may have stepped in to run Addis Ababa Café to give his aging and ill mother a much-needed break, but it's still Yenenesh who sets the tone for the restaurant, even though she's no longer a regular presence inside its walls. Zed admits he's not as tough about it as his Mum. He still shows visitors where to wash their hands in the small basin in a screened-off corner, in preparation to tear at the injera—an Ethiopian flat bread—with their hands to scoop their food, but knows he's "not as particular." It's "her age group," he explains, "[and] their way of communicating." Zed doesn't so much rebel against this approach as reformulate it: "Growing up here, I know how to get the message across in a different way."

However much Zed's own experience has changed his method of communication—spending most of his childhood in suburban Australia, whereas Yenenesh grew up in Ethiopia—Yenenesh's vision still defines what Addis Ababa Café *is*. In reviews, Yenenesh is likely to be noted by her absence. Rachel Lebihan, for example, points out that Zed is "holding the fort for his mother and family matriarch, Yenenesh" (Lebihan 2010). Zed also acknowledges that the changes he's introduced while running Addis Ababa Café, though large, are ultimately cosmetic: "She's the heart and the guts of everything—I just make it look pretty."

Through Zed, nevertheless, Yenenesh's conception of Addis Ababa Café is reformulated, but the core remains unchanged. Yenenesh describes one part of her motivation: "I loved my culture [and wanted] to show [it to the] Australian people." (English remains the main spoken language in the restaurant, but Amharic is spoken with Ethiopian customers.) Zed expands on this idea, noting that "she never really built it for profit—she's always wanted to do catering, she's always wanted an expression of the food in the culture."

The key to Zed's approach of making "pretty" the core of experience and presentation that Yenenesh provides—of "exploiting anew" (Hountondji 1996, 161) his mother's and, in many ways, his own cultural history—has been, in recent times, through online networking on Facebook. Facebook has become a digital space that both repeats and recomplexifies the tensions and difficulties inherent to Addis Ababa Café's offline geography and multi-generational communications. The desire to represent a "home away from home" in a new location far from its "roots," and to preserve yet stretch the boundaries of "tradition" across dimensions of age, ethnicity, class, dietary preferences, religion, and so on, becomes as much a digital problem as one of the offline everyday.

Facebook on the Menu: Hybrid Familial Networks

Addis Ababa Café's appearance on Facebook was, in fact, encouraged by the restaurant's loose sense of offline familial networking: specifically, by the café's positive reviews in local papers and food guides, as well as by enthusiastic customer word-of-mouth advertising. One night, a regular customer asked why the restaurant wasn't trying to capitalize on its vegetarian options, suggesting the possibility of special vegetarian "banquet" nights. These banquets are now advertised widely through the restaurant's Facebook group and email lists, as well becoming word-of-mouth events for restaurant regulars; they have become an established feature of Addis Ababa Café's identity, replacing nonvegetarian banquets entirely. Though Zed has tried, and continues to try, various forms of online promotion—such as a Youtube promotional video and online "voucher" sites such as LivingSocial.com—Facebook has proven particularly important for Addis Ababa Café's growth, both in its ability to establish a network based upon community building and its capacity for overt displays of "ethnic" engagement.

In other words, Addis Ababa Café's networks are hybrid, with the café providing both a taste of "home" for the "local" Ethiopian community and a taste of "elsewhere" and the "exotic" (or perhaps, simply cheaper, homely, richly flavored food) for the "local others." In the ingredients for injera and doro wet (spicy chicken stew), then, lies the possibility of a complex range of nostalgic memories and imaginings.

Though Addis Ababa Café has appeared in a number of media publications, often with an "ethnic" slant to the story, an extended advertising presence in the media gave few results, as Zed explains:

For advertising like that you'd pay hundreds of dollars and it probably wouldn't work. . . . I thought I'd pursue a . . . webpage . . . [one that] tells you where we are and all that stuff, then I thought no one's gonna go through that. So I thought "they spend $400 for a little write up, and then there's nothing." And people actually do it through Facebook, and it does heaps better.

The ongoing process of Addis Ababa Café's exposure on Facebook has become a natural part of the restaurant's offline identity, even though there is little in the restaurant's warm, homey, and nostalgic appearance that signals this offline-to-online growth. Online information gathering has remained intrinsically linked to offline space: email addresses are still gathered by the traditional guest book that sits at the restaurant's front counter. Documenting the daily life of Addis Ababa Café is now a part of Zed's regular routine: "I use my camera so much that I broke one. I always carry a camera with me, and the whole process of carrying it everywhere damaged one and I had to get another one."

In fact, contrary to usual practice, the idea of the guest book emerged from Addis Ababa Café's increasing move toward online networking after the creation of their Facebook group, rather than the other way around. The new media networking provoked the return to a traditional form of gathering information and cataloguing and displaying experience.

Facebook has acted as an extension of Addis Ababa Café's growth based on "word of mouth . . . at the beginning," says Zed, noting that "technology makes a change." Experience at Addis Ababa Café is understood as an interactive and communicative process, rather than a one-way delivery:

They post comments . . . they've been here and they enjoyed their experience. . . . There are a lot of reviews . . . about the café. People tell me, so I go on the sites and I download them and I put them on Facebook . . . they just send a comment, then they forward it to a friend . . . great to just communicate. . . . It helps, I mean, it does, whenever I do any functions, that's how I do it, I just put it on Facebook and I email the data that I created on Facebook. See, Mum never used to have that.

An African Hut on the Highway: Transformation from Within

In many ways, the physical emergence of Addis Ababa Café in West Hindmarsh is replayed by its digital emergence on Facebook. Although the café was built up from within an existing and seemingly predetermined physical structure, it finds its own personal resonance and reaches out from within. On first encounter, Addis Ababa Café is an African hut, nestled somewhat incongruously behind a chicken shop, in a fairly drab industrial/suburban area on the car-heavy but people-sparse Port Road. Visible from outside the hut is a large office supplies store, positioned directly opposite on the multilane pedestrian-unfriendly highway. The hut is composed of bamboo strips, the veneer of an "authentic" African café draped over and obscuring the suburban regularity of the 1970s-style commercial structure. The "authentic" hut facade was constructed with the aid of local friends from supplies purchased at a popular national hardware chain.

Council approval, usually required for major changes to buildings, was easily avoided: the transformation from suburban shop to African hut was effected entirely within the confines of the existing structure. The sheets of wood disguise the rigid, square redbrick frame, but do not displace it. The roof is still visible above the facade and the redbrick pillars still delineate the edges, but the entryway now suggests a hint of an alternate space in the seemingly monocultural industrial expanse, whereas previously, Zed describes, "there was nothing there, when you close this door you wouldn't even realize it's a restaurant.

Like Addis Ababa's facade before its transformation, Facebook's unyielding uniformity can seem to reassert the monocultural expanse of the industrial suburbs. Certainly, Facebook's rigid style and presentation appears to dissuade rather than encourage possibilities of transformation and the expression of the aesthetically and culturally "different." Such an understanding of the online realm thus leaves the online space and community fostered by Addis Ababa Café open to the same considerations as the offline world when it comes to constructing and preserving the presence of "difference." What is it that draws membership to the Addis Ababa Café Facebook page?[2] How is a sense of community fostered around this alternate space, and is the engagement really a true exchange of cultures or merely a cultural browsing that leaves both parties more or less untouched? Is this the equivalent of visiting a food market as a tourist in search of gastronomic adventures (Duruz 2004)?

Online Ethnic Grazing and Browsing

If Facebook and the online paradigm are defined as a suburban-style retreat (Boyd 2010), then Addis Ababa Café's presence on the site may suggest not only a portrait of a community founded in Ethiopian identity showcased in a primarily "white" space, but also the social activist bell hooks's sense of a space where "ethnicity becomes spice, seasoning that can liven up the dull dish that is mainstream white culture" (hooks 1992, 21). Similarly, Ghassan Hage presents such browsing as a cosmo-multiculturalist fantasy: "He [the white cosmopolitan] wants them [ethnic providers] to 'cook-for-themselves-in-community' *for him*" (Hage 1997, 140; see also Duruz 2004). Hage develops this argument by criticizing the politics underlying performances of "ethnic" cultures—exhibitions of food, dance, and music at festivals that "enrich" mainstream culture without intervening in overall power structures:

> Far from putting "migrant cultures," even in their "soft" sense (i.e., through food, dance, etc.), on an equal footing with the dominant culture, the theme conjures images of a multicultural fair where the various stalls of neatly positioned migrant cultures are exhibited and where the real Australians, bearers of the White nation and positioned in the central role of the touring subjects, walk around and enrich themselves. (Hage 1998, 118; see also Antonio 2001, 370)

Certainly, recent commentators have stripped the online realm of much of its utopian community gloss; the same kinds of tension are found in a realm that once offered to liberate us from them. The status of online spaces remains as ambiguous, problematic, and in need of regular and ongoing analysis as traditional offline spaces. It is, of course, impossible to consider one without the other. Although, "spectacular flights of fantasy" (Robins 1995, 136) sometimes still emerge: imaginings of an online realm with the capacity to transcend cultural and political reality.

Further reaffirmations in Facebook and online networking of real-world ideology are also suggested by Danah Boyd (2010), who presents Facebook as an "elite" and "white" realm, and part of a race-based self-segregation of online social networking users. Boyd compares the move from MySpace to Facebook with "suburban white flight": "Facebook's origin as a gated community and parents' belief that the site is private and highly monitored reflect the same values signaled by the suburbs" (Boyd 2010, 34). Far from being a disruption, electronic advances simply allowed new locations for the existing hegemony: "Structurally, social networks are driven by homophily even when there are individual exceptions. And sure enough, in the digital world, we see this manifested right before our eyes" (Boyd 2009).

Paul C. Adams and Rina Ghose agree, issuing a warning in their study of Indian "bridgespace": "Society must be understood in terms of power imbalances. New technologies play a clear role in accentuating social divides and favoring pre-existing authority structures" (Adams and Ghose 2003, 416). In summary, we can turn to the cultural theorist Slavoj Žižek who argues that simple knee-jerk immersion in the "new paradigm" remains a "deeply conservative gesture" (Žižek 2001) and evades rather than addresses the already-existing concerns and tensions inherent in the supposedly new realm.

As such, it is vital to focus not on the "new" itself (the online realm), but on how the "new" replicates and reformulates the existing complexities of the "old": how, as Matthew Zook says, the online environment "provides new geographies of connection and exclusion (e.g., the difference in access available in physically proximate downtown business districts versus downtown slums); invites experimentation; and opens the possibility for contests between differing visions of the world" (Zook 2006, 53).

While an online community like Facebook can be seen as a fundamentally reaction-ary environment, "organized by race" (Boyd 2010, 16) and defined by an extension of real-world racial segregation and hegemonic structures, there nevertheless remains the opportunity for these spaces to be a place for building alternate and transformative communities. As we suggested in our introduction, if the online realm reflects the underlying concerns of everyday life, then this also includes the benefits and the problems, the issues and debates, relating to the ongoing cultural project of living "together-in-difference" (Ang 2001).

"Liking" as Cosmopolitan Performance: Identity Construction Offline and Online

Not only does the conception of "touring subjects" extend to breezily navigable online spaces, but it also becomes particularly potent when online identity on Facebook can be so easily and superficially constructed and reconstructed through a network of surface-level presentations of association. The commodifying underpinnings of iden-tity construction may be further accentuated by the fact that "Facebook users predomi-nantly claim their identities implicitly rather than explicitly; they 'show rather than tell' and stress group and consumer identities over personally narrated ones" (Zhao, Grasmuck, and Martin 2008, 1816). Furthermore, "the emergent nonymous Facebook environment provides users with new leverage for selective self-presentation" (Zhao, Grasmuck, and Martin 2008, 1821), and "consumer/taste identity [is] defined as much by what the market offers as by individual or character traits" (Zhao, Grasmuck, and Martin 2008, 1826).

Indeed, Facebook self-presentation can frequently represent itself primarily through displays of connection with groups and predefined preferences rather than specific "unique" representation. As Shanyang Zhao and his colleagues suggest, presentation may take the form of "see me first and foremost in the context of my group," with some users spending "more time defining what cultural niche they are part of than staking a claim of individuality distinct from these desires/tastes" (Zhao, Grasmuck, and Martin 2008, 1826). Subcultures thus may not be granted the freedom and power initially perceived as part of the online environment. After all, when these are specifi-cally tied to the primarily "nonymous" environment of Facebook, they are constrained by "the presence of social pressure and a degree of censorship on Facebook, which is not something commonly experienced in anonymous online settings" (Zhao, Grasmuck, and Martin 2008, 1830).

In fact, it can be argued that the "ethnic" *provider* of identity associations in this context—such as Addis Ababa Café's Facebook page—scarcely need *perform* at all. Rather, the provider merely needs to be *available* so that association (joining the group, "friending," or "liking") can be made visible and "public." If identity is largely a performance, and this is heightened in the Facebook environment, then it might seem that the performance of the provider is secondary to the performance of the cosmopolitan browser: it is the cosmopolitan who, in fact, performs by "liking" the mere presence of the provider.

Alternate space is thus made more available, but potentially with less opportunity to demonstrate the very points of difference through which it may seek to define itself. The act of "liking" may simply assert, with uncomfortable transparency, the redundancy of the "ethnic" performer in this process.

Others Constructed by Others: Associative Identity and Facebook Nostalgia

As a result of this element of identity-by-association, the presentation of "self"—whether by a person, business, or cultural identity—is not governed solely, or even primarily, by a user's own immediate manipulation of their profile, even if such control is open and available to them. In fact, more weight may be placed on the defining features external to the subject, as "in a Facebook profile, things that others say about a target may be more compelling than things an individual says about his- or herself. It has more warrant because it is not as controllable by the target, that is, it is more costly to fake" (Walther et al. 2008, 33).

If we accept that "behavioural residue left by their Facebook friends can be construed as sanctioned, at least in the eyes of observers" (Walther et al. 2008, 35), then Addis Ababa Café's online identity is thus an interaction between its own sense of identity, and the identity that is thrust back upon it by those who associate with it most openly. Facebook "wall" analysis thus plays an important role in how we construct a Facebook identity. In the study of Joseph Walther and his colleagues, for example, "results showed that complimentary, prosocial statements by friends about profile owners improved the profile owner's social and task attractiveness, as well as the target's credibility" (Walther et al. 2008, 44). The notion that people "make different kinds of spaces in technologies" (Adams and Ghose 2003, 433) thus cannot be limited to those who present themselves online, but must be seen as a broader creation by all those who engage with this self-presentation.

The construction of Addis Ababa Café—in line with Yenenesh's vision of a family or cultural center where people can engage with community and "nostalgia" rather than simply purchase food—is not accentuated by postings of Zed and Mel [Zed's partner and co-owner of the business] themselves (who rarely, if ever, post status updates not related to specific restaurant events), but is in fact more definitively propagated by general customers and Facebook users. The image is primarily constructed by other users who are not themselves directly responsible for the representation of Addis Ababa Café and its cultural values in the online realm.

Recent Facebook posts on Addis Ababa Café's (easily accessed and shared) online profile emphasize features such as authenticity, family interaction, and community (user names have been altered, but all quotes retain original spelling and grammar):

July 27, 2010, Sarah: "Authentic Ethiopian Food"

The idea of discovery and transmission of the experience also features:

March 13, 2010, Jillian: "when is there going to be another banquet night??? i want to enlighten my friends to the delish food you have"

Similarly, there is no shortage of posts thanking Mel and Zed directly as though having had dinner with family friends:

March 9, 2009, Robert: "Hi! We had a DELICIOUS meal with you guys tonight. It was lovely to meet your children, thank you very much for your great food and hospitality. We'll definitely be back!"

November 16, 2009, Beth: "Hey guys thanks for another geat meal on Friday night!! Tariqua had the biggest smile on her face when she saw the injera. Was good to see you were so busy. Keep up the great work. We'll be back soon!!"

December 5, 2009, Meg: "Thanks for the great meal tonight Zed, and thanks for taking the time to chat with us:)"

January 24, 2011, Margaret: "What an amazing resturant I am so Glad that I have meet such wonderful and talented people, with the most incredible food. Thank you Mel and Zed for such a wonderful resturant we will be all seeing you again soon"

February 2, 2011, Linda: "You guys are awesome!! Thanks for always looking after us when we come visit you and feed our faces:) always go home sooo full xx"

As a locus of public (and peer) display, this emphasis on experience and authenticity acts not only as praise solely for the benefit of Addis Ababa Café itself. These messages also function as a point of identification for those who believe "that crossing cultural boundaries easily [is] a positive characteristic," (Antonio 2004, 563); Anthony Antonio also notes (in a study of students' racial and social ties) that "these values have led them to cross cultural boundaries frequently and to choose friends who hold similar values" (Antonio 2004, 563). The benefits of such displays of inclusion may also be interpreted as further cultural "browsing" on an institutionalized level, where association with alternate cultures is valued over extended interaction and serves to increase social capital among a "browsing" community.

As such, the community building around a digital presence such as Addis Ababa Café's may be a construction of ethnic identity largely out of the "ethnic" performer's hands. When asked about the growth of the Facebook page and whether or not they actively sought group members to join, Zed seems mostly amused by the lack of personal involvement required: "We just left it!"

Connecting with Alternate Communities: Vegetarians and Vegans at Addis Ababa Café

In the midst of these complexities of identity and association, of "checking up on regularly" (Joinson 2008, 1035) and "surveillance" as ideological regulation (Lampe, Ellison, and Steinfield 2007, 167), one of the unexpected cultural collaborations brought about by Addis Ababa Café's presence on Facebook has been with the vegetarian and vegan community. This point of collaboration has extended Addis Ababa Café's connection with vegetarianism and the adoption of a vegetarian identity from a casual afterthought to something so significant that "we'll never cancel."

In line with the changing and ever-developing nature of African identity suggested by Paulin Hountondji (1996), vegetarianism emerges from Addis Ababa Café's Ethiopian identity as a cultural assemblage that is both traditional and modern, authentic and evolutionary; "African" and "vegetarian" are two niche cultures that perhaps, in a popular understanding in the West at least, have only superficial connection.

Addis Ababa Café's growth as a vegetarian destination was primarily a fortuitous event of cultural synchronicity, with vegetarianism an element underlying much of Ethiopian culture that related to fasting periods where no animal products are consumed. Although previously unexamined by Zed as part of his cultural background, vegetarianism emerged as something available to be explored and exploited as part of Addis Ababa Café's cultural identity: only, however, after confirmation of its "authenticity" by Yenenesh. As Zed explains of the fortuitous synchronicity:

We do have regular customers who come here for vegetarian, and I started normal banquet nights . . . and the vegetarian group were like, "it's only three choices on the menu, we want something like this for us too." So I thought, "OK, maybe I can put together a banquet." So I spoke to Mum, "What are all the different vegetarian foods we can do in our culture . . . and how many are there?" And she said, "We probably have more vegetarian dishes than meat dishes" . . . because we have a fasting period for fifty-two days of the year, they fast in Ethiopia, which is all [food with] no animal products in them. . . . It's a cultural, religious, Coptic Orthodox . . . fasting before Easter, it is very much like that, fifty-two days of the year.

Importantly, Zed ties this expansion of Addis Ababa Café's vegetarian menu directly to the initial creation of the Facebook group:

She [Yenenesh] gave me different names, and I thought OK, we'll do about eight different ones and try them out, and Mel has a Facebook of her own and I thought, can you create one just for the café and see how many people join, and then we can start to communicate to them through that, and in a matter of a week about a hundred people joined. . . .

The banquet night was pretty much birthed out of that. . . . Usually about 85 is the most I can sit down. . . . Every six weeks we've been doing them. They're only vegetarian now. . . . that's the popular group, they want them. By demand! They love the fact . . . they write emails too about their experiences . . . a lot of reviews. . . . And I put it on Facebook. . . . [There was] so much [vegetarian] demand. . . . Animal Liberation Association put us on their email list . . . first few times, just from that group, about twenty-six people came.

This unanticipated "claiming" by vegetarian groups (as well as animal liberation groups) has been key to Zed being able to "exploiting anew" (Hountondji 1996, 161) his cultural traditions, while adhering to "authenticity" through Yenenesh's validation of vegetarian Ethiopian culture. Advertised through Facebook and email alone, the vegetarian banquets are frequently booked solid within a week of being announced, with most customers drawn by the prospect of tasty vegetarian/vegan dishes and having little or no experience with Ethiopian food (the banquets offering exposure to a variety of dishes at one time). As most of Addis Ababa Café's vegetarian Ethiopian food is "vegan anyway" ("no animal product at all—they love that"), the provision of a space for vegetarian culture has also extended naturally to vegan customers and cultures, transforming the nature and conception of Ethiopian culture, yet without challenging the facticity of its presentation and its meanings of authenticity. As well as on banquet nights, additional vegan options have become firmly entrenched as part of the everyday menu.

This kind of naturally (and fortuitously) occurring synchronicity between "alternate" communities from "different backgrounds and experiences of life" and spaces can also be seen in Addis Ababa Café's increasing appeal for an Asian clientele:

A lot of interest among Asian people . . . there's a big group in church, [an] Asian group, that came here, and they tend to email each other too. . . . They like a traditional meal too, to sit down in a traditional way, and spices too. On Saturday, I have about fourteen Singaporeans coming and they love this place, one of them, a doctor, tried it out . . . the chicken especially, they don't eat a lot of beef but they like chicken. So I said, "Don't worry I'll do you the chicken all you like!" And the vegetarian spicy dish, they love that, they really like the lentils, so they're into the spices.

Linking to Origins: Bridgespace and the Local

While this kind of "ethnic" online presence can be used to engage with others who identify personally with this specific representation of cultural authenticity, the expansion and growth that merges with other surrounding identities is also a key part of this online and offline composite identity building. For Adams and Ghose (2003), the use of disparate technological communities to reconnect with a fundamental sense of ethnic identity—a concept that "implies a place of origin" (414)—presents an interesting irony. Something "new" (technology) enables the re-experience of something "old" or, in their words, "a new technology can be used to cognitively connect with what is a symbol of primordial essence" (Adams and Ghose 2003, 415). They explore this through a notion of "bridgespace"—"a life built upon the Internet's facilitation of creating roots in multiple places, thus generating 'different types of space in technologies'" (Zook 2006, 63). In this reckoning, bridgespace acts as a facilitator of cultural movements and exchanges depending on how it is used and established by individual participants: "a collection of interconnected virtual places that support people's movement between two regions or countries and the sustenance of cultural ties at a distance" (Adams and Ghose 2003, 419).

For our own purposes, it's interesting to adjust this notion of "cultural ties at a distance." The ties discussed above in relation to Addis Ababa's engagement with local communities and groups are as much about Australia—and the local—as they are about Ethiopia or a "place of origin" (Adams and Ghose 2003, 414). The "symbol of primordial essence" (415) on offer perhaps allows that sense of "ethnicity" to be less a "real" time than an "imagined time when a 'people' lived together in a particular place" (414). Addis Ababa offers a bridgespace not to those who have a direct identification with Ethiopian ethnicity, but to those who choose to associate with it as part of a local culture and—perhaps more intriguingly and as we have seen above—those who find themselves in unexpected synchronicity.

As Yenenesh points out, her spices are simultaneously Australian and Ethiopian, an "ethnicity" constructed from within the "local," her ingredients sourced from local markets as replacements for unavailable Ethiopian ingredients (for discussions of this kind of "localization" see also Tan 2007, 181). The bridgespace is offered up for those who might have no access to it otherwise (and potentially no apparent *reason* to access it), believing it to be external to their own culture rather than an element within it. In this way, it is possible to be nostalgic in a number of senses: for tastes, smells, and textures that one has lost or for the ones that we have never owned in the first place, but yearn to own.

The role of "innovation" (Adams and Ghose 2003, 430) in bridgespace further suggests that Yenenesh offers a bridgespace as much for the local as for the "ethnic." She provides not merely a "tension between tradition and innovation," but a reaffirmation of the co-existence—the requirement—of "tradition and innovation" as part of her own "doing-cooking" (Giard 1998, 149), cooking-as-memory, and local community building through "authentic" identity.

Conclusion: The Tension between "Tradition and Innovation" in Addis Ababa Café

For Addis Ababa Café, the seemingly organic transition to online social networking through Facebook has served not to facilitate a simple representation of a more "authentic" Ethiopian presence and locus for a cultural nostalgia, but to render cultural resonances more complexly and provoke an innovation in the representation and exploitation of cultural identity, in both online and offline spaces. Facebook has, in a sense, accentuated the bridgespace role of Yenenesh's desire for community engagement with her sense of the importance of her own cultural history, where food acts as experience and facilitator of memory. The bridgespace, in a sense, acts as a mediator between the "hut" of Addis Ababa Café's exterior—with its simultaneous connotations of the exotic and the familial—and the seemingly monocultural expanses of both the suburbs, and of Facebook itself; it links not to a lost culture, but back into its own local culture, transforming it and allowing itself to be transformed.

By stepping into the online realm, Zed has embraced his position as a cultural intermediary (though it is his own culture and history for which he acts as messenger), "performing" for a new kind of online cosmopolitanism, but also being "acted upon" as new affiliations and synchronicities tease out new meanings and memories within Ethiopian culture. Facebook continues to play a key role in Addis Ababa Café's self-presentation and community engagement; a recent appearance by Zed at a WOMAD-elaide 2012 music festival cooking show saw a surge in business and Facebook visitors and members. Through Facebook, an alliance has been negotiated between an "authentic" Ethiopian culture and vegetarian and vegan communities, transforming not only the popular understanding of Addis Ababa Café's cultural resonances, but also Zed's own understanding of his own Ethiopian heritage.

As Ien Ang points out, cultural identity is always in a state of "in-betweenness" (2001, 194); the development of Addis Ababa Café's Ethiopian identity through online engagement and affiliations suggests complementary values of "authenticity" and new "exploitations." These ongoing manifestations of a "complex heritage" and the notions that "African civilization" is not "a closed system in which we may imprison ourselves (or allow ourselves to be imprisoned)" (Hountondji 1996, 161) is integral, and inescapable, in the bridgespace that Yenenesh offers and that Zed, through Facebook, has expanded into new communities and synchronicities.

For Yenenesh, such "innovation" has been her natural understanding of cooking all her life. When asked about her food's origin, she readily locates it in her new environment, claiming, rather than downplaying, the necessary adjustments she has had to make to compensate for the lack of expected ingredients and equipment in her new home. "From here. From Australia," as she says of her Ethiopian spice.

VEGETARIAN FOSOLIA (ETHIOPIAN STEW)

Makes 4 servings

This recipe is adapted from Zed Wondimu of Addis Ababa Café, who included the original in WOMADelaide 2012 festival booklet, "Taste the World: Recipes."

Vegetarian fosolia (accompanied by tibs, Ethiopian-style sauteed, marinated meat for nonvegetarians) is usually eaten with injera, the traditional Ethiopian bread. Injera, a unique, spongy, yeast-based pancake-like bread, is meant to be rolled up and dipped into stews and sauces, and used as such in place of cutlery.

Ingredients

1 lb. frozen cut green beans

1 lb. frozen julienne carrots

2 large brown onions

2 tbsp. minced garlic

2 tbsp. minced ginger

¼ cup vegetable oil

Salt to taste

1 tbsp. white sugar

1 cup water

Preparation

1. In a pot, combine the frozen carrots, beans, onion, oil, and water and steam until soft.
2. Reduce the heat to medium and add the garlic, ginger, sugar, and salt. Stir and cook 5–7 minutes with the lid on. Be sure not to let it dry out or else the base will burn! Keep it moist.

Notes

1. The Australian Department of Immigration "identified approximately 3,000 Ethiopian people settling in Australia" between 2000 and 2005 (Commonwealth of Australia 2006, 4), with some 3,600 Ethiopian-born people living in Australia according to the 2001 Census. Of the 2000–2005 arrivals, only 249 settled in South Australia (with most Ethiopians settling in Victoria, a recorded number of 1,546). The 2006 Census data show 5,634 respondents listing Ethiopia as country of

birth, with under 500 residing in South Australia (Australian Bureau of Statistics 2008). The restaurant's council area (Charles Sturt) does not include Ethiopia or African countries in its list of main birthplaces, placing Ethiopian origin as well below than the lowest listed percentage of 0.9% of local population (Charles Sturt Council 2007, 5).

2. The Facebook page has 339 members at the time of writing, rising, and falling over time and with Facebook format changes.

References

Adams, Paul C., and Rina Ghose. 2003. India.com: The construction of a space between. *Progress in Human Geography* 27 (4):414–437.

Anderson, Benedict. 1983. *Imagined communities*. London: Verso.

Ang, Ien. 2001. *On not speaking Chinese: Living between Asia and the West*. London: Routledge.

Antonio, Anthony Lising. 2001. Diversity and the influence of friendship groups in college. *Review of Higher Education* 25 (1):63–89.

Antonio, Anthony Lising. 2004. When does race matter in college friendships? Exploring men's diverse and homogeneous friendship groups. *Review of Higher Education* 27 (4):553–575.

Australian Bureau of Statistics. 2008. Census 2006—people born in Africa. *3416.0—Perspectives onmigrants,2008*.http://www.abs.gov.au/AUSSTATS/abs@.nsf/Lookup/3416.0Main+Features32008.

Boyd, Danah. 2009. The not-so-hidden politics of class online. *Personal Democracy Forum*, New York, June 30. http://www.danah.org/papers/talks/PDF2009.html.

Boyd, Danah. 2010. White flight in networked publics? How race and class shaped American teen engagement with MySpace and Facebook [draft]. In *Digital race anthology*, ed. Lisa Nakamura and Peter Chow-White. Routledge Press [forthcoming]. http://www.danah.org/papers/2009/WhiteFlightDraft3.pdf.

Charles Sturt Council. 2007. "Charles Sturt Council snapshot." http://www.charlessturt.sa.gov.au/webdata/resources/files/Council_Snapshot.pdf.

Commonwealth of Australia. 2006. Ethiopian community profile. Australian Government Department of Immigration and Citizenship. http://www.immi.gov.au/living-in-australia/delivering-assistance/government-programs/settlement-planning/_pdf/community-profile-ethiopia.pdf.

Duruz, Jean. 2004. Adventuring and belonging: An appetite for markets. *Space and Culture* 7 (4):427–445.

Giard, Luce. 1998. Doing-Cooking. In *The Practice of Everyday Life, Volume 2: Living and Cooking*, ed. Michel de Certeau, Luce Giard, and Pierre Mayol, 152–159. Minneapolis: University of Minnesota Press.

Hage, Ghassan. 1997. At home in the entrails of the West: Multiculturalism, ethnic food and migrant home-building. In *Home/World: Space, community and marginality in Sydney's West*, ed. Helen Grace, Ghassan Hage, Lesley Johnson, Julie Langsworth, and Michael Symonds, 99–153. Annandale, NSW: Pluto Press.

Hage, Ghassan. 1998. *White nation: Fantasies of white supremacy in a multicultural society*. Annandale, NSW: Pluto Press Australia.

hooks, bell. 1992. Eating the other. In *Black looks: Race and representation*, ed. bell hooks, 21–39. Boston: South End.

Hountondji, Paulin J. 1996. *African philosophy: Myth and reality*. Henri Evans. trans., with the collaboration of Jonathan Rée. Bloomington, Ind: Indiana University Press.

Joinson, Adam N. 2008. "'Looking at,' 'looking up' or 'keeping up with' people? Motives and uses of Facebook." In *CHI '08 Proceedings of the Twenty-Sixth Annual SIGCHI Conference on Human Factors in Computing Systems*, 1027–1036. New York: ACM New York.

Lampe, Ciff, Nicole B. Ellison, and Charles Steinfield. 2007. The benefits of Facebook "friends:" Social capital and college students' use of online social network sites. *Journal of Computer-Mediated Communication* 12 (4):1143–1168.

Lebihan, Rachel. 2010. Spoilt for choice. *The Australian Financial Review,* December 29, 2009–January 3, 2010, L7.

Robins, Kevin. 1995. Cyberspace and the world we live in. In *Cyberspace, cyberbodies, cyberpunk: Cultures of technological embodiment*, ed. Mike Featherstone and Roger Burrows, 135–155. London: Sage.

Tan, Chee-Beng. 2007. Nyonya cuisine: Chinese, non-Chinese and the making of a famous cuisine in Southeast Asia. In *Food and foodways in Asia: Resource, tradition and cooking*, ed. Sidney C.H. Cheung and Tan Chee-Beng, 171–182. London: Routledge.

Walther, Joseph B., Brandon Van Der Heide, Sang-Yeon Kim, David Westerman, and Stephanie Tom Tong. 2008. The role of friends' appearance and behaviour on evaluations of individuals on Facebook: Are we known by the company we keep? *Human Communication Research* 34:28–49.

Zhao, Shanyang, Sherri Grasmuck, and Jason Martin. 2008. Identity construction on Facebook: Digital empowerment in anchored relationships. *Computers in Human Behavior* 24 (5): 1816–1836.

Žižek, Slavoj. 2001. The one measure of true love is: You can insult the other. Interview by Sabine Reul and Thomas Deichman. *Spiked* November 15. Available at: http://www.lacan.com/zizek -measure.htm

Žižek, Slavoj. 2004. *The reality of the virtual*. Directed by Ben Wright. Ben Wright Productions.

Zook, Matthew. 2006. The geographies of the Internet. *Annual Review of Information Science & Technology* 40 (1):53–78.

3 What Are We Going to Eat Today? Food Recommendations Made Easy and Healthy

Jettie Hoonhout, Nina Gros, Gijs Geleijnse, Peggy Nachtigall, and Aart van Halteren

For several decades now, food has been "omnipresent" and easily available, in shops and other food outlets, in many parts of the world. Choice has increased as well in recent decades: for example, the once-simple process of ordering coffee or a sandwich now requires making a selection among a wide range of options and extras on offer. In addition, the amount of information about good versus bad nutrition available to the public has increased enormously. Government organizations, the food industry, healthcare professionals, and scientists, all try to inform and thus influence consumers in their choices. Numerous campaigns may offer nutritional advice to the public about general issues (e.g., eat more fiber; consume less salt; eat more fish, vegetables, and fruit). Publications in the popular press (such as Baer, 2010) promote the benefits of eating specific foods : yogurts that improve bowl function, tomatoes that are good for your skin, broccoli that protects against cancer. This large amount of diverse information is for many consumers quite confusing and complex, making it difficult to decide what is fad and what is wisdom.

A Problem with Food?

Newspaper stories about the rising numbers of overweight people might give us the impression that food itself has become a problem. In recent years, the eating habits of people in Western countries—and increasingly in other parts of the world as well—have changed in many ways, often with dramatic results. Maintaining a healthy weight is a growing problem among adults, and increasingly also among children. The estimated prevalence in Europe of overweight among children is 30 percent (Eurobarometer 2006). The latest projections by the World Health Organization (WHO 2006) revealed that 1.6 billion adults aged fifteen and older were overweight and at least 400 million adults were obese. For 2015 these numbers are expected to increase to 2.3 billion overweight and 700 million obese adults (WHO 2006). Obesity-related illnesses are estimated to account for 7 percent of the total healthcare costs in the European Union alone (Branca et al. 2007).

These statistics have inspired many to call for interventions through weight management programs, or via education, advertising, tax and price policies, and so on (Nestle 2003). However, treating weight problems appears to be extremely difficult. While diets often do result in short-term effects, the maintenance of bodyweight loss in the longer term is often unsuccessful (Miller 2001; Paradis and Cabanac 2008; Riebe et al. 2004).

Prevention, then, seems to be a more advisable strategy. But current efforts put in prevention programs and campaigns, often in the form of generally worded recommendations advocating a healthier lifestyle (e.g., eat more whole grains, eat less saturated fat), often have little effect. One reason is that factors such as availability and access to (healthy) food are actually making it hard to follow such advice (Branca et al. 2007). Another reason is that such preventative campaigns do not provide personalized feedback that takes into account people's preferences and their physical, mental, and environmental situation. In fact, it is more likely that the amount of information overburdens human cognitive capacities to use it appropriately during food-choice processes (Häubl and Murray 2006; Walls et al. 2009), even for motivated consumers (Contento 2007). This glut of information might also lead to demotivation ("this is too complex for me; there is nothing I can do about it"). The huge amount of offerings in the food market combined with the ever-increasing amount of often ambiguous nutrition information can easily cause difficulties for overwhelmed consumers unable to decide which option is best or "right." We could expect that perhaps, given all these problems, people's interest in healthy food would be low. And yet research suggests this is not the case (Barrier 2005; Contento 2007; Rosin 2008).

Can Computers Help?

Computer applications are gradually becoming more common in the world of health, nutrition, and diet. Some researchers see computer systems as holding potential in nutrition education (Brug, Oenema, and Campbell 2003). Others (Kraft, Drozd, and Olsen 2008) argue that digital interventions might be helpful to foster sustained behavior change. Pei Yu Chi and colleagues (2008) have presented an application to provide feedback during the cooking process about the amount of calories in the dishes people prepare, which is thought to have a positive influence on the confidence and willingness needed to cook healthy meals. Web services offering tools to log food consumption in an online food diary are now common. For example, the website of the Medical Research Council provides dietary assessment tools[1] and MyFoodDiary.com includes an online calorie counter.[2] Unfortunately, the research in the area of computer systems supporting a healthy diet rarely investigates the effectiveness and acceptability of these kinds of interventions in the long term.

In this chapter we will discuss how users can be guided with the help of computer technology to adopt a healthier lifestyle. We will present the development of a tool that provides actionable guidance in making healthier choices in meals, in terms of calories and nutrition. We have organized the chapter into the following sections: an introduction to the recipe application; the test design we used to research its effectiveness; the findings of the evaluation; and our conclusions.

Toward the Development of a Support Tool

The literature on nutrition and health (e.g., Branca et al. 2007; Contento 2007; Mojet et al. 2001) suggests a range of issues that influence consumer food choices, including individual characteristics, household composition and family situations, time constraints, social conventions and constraints, and cultural issues. In addition, we talked to many consumers about their struggles with health and food—their opinions, beliefs, perceived barriers, and current practices—to get a feeling for the issues they experience and help us to generate ideas for how technology might address these issues. A detailed presentation of the findings from the literature and our discussions with consumers is beyond the scope of this chapter. But we used our findings and collected insights in a series of brainstorming sessions with colleagues.These brainstorming sessions resulted in a number of ideas for how people like the consumers we talked to might be helped. A key issue had emerged from our research: although our intended target group (people who would like to eat healthier than they do now, and who perceive themselves in principle as capable of preparing healthy meals) seems motivated, for many it is difficult and cumbersome to implement this desire in practice. In other words, people's intention to eat in a healthy manner does not always lead to corresponding behavior. In most cases, other activities interfere and certain barriers, not always entirely under their control, keep them from implementing their intentions. This discrepancy between intention and behavior can be described as a "gap." Icek Ajzen, Cornelia Czasch, and Michael G. Flood (2009) argue that a simple intervention, such as the formulation of a specific, concrete plan, can help to decrease this discrepancy and thus reduce the "intention-behavior gap." We designed our concept in response to this need to support the formulation and execution of a concrete plan of action.

Detailing and Building the Prototype
Many consumers are interested in learning more about the nutritional value of a meal, but find it difficult to calculate or even estimate this value (calories, saturated and unsaturated fats, proteins, vitamins, fibers, etc.) of a cooked meal consisting of different and varying amounts of ingredients. So, we decided to address this task in particular. Following the outcome of the brainstorm sessions, we decided that a *recipe tool*

would be an interesting platform to further investigate how to support our target group. Using software technology (see Geleijnse et al. 2010, for more details), we built a recipe browser and cooking guide, called MyCookingCompanion. The application was designed to provide an easy way to select daily meals without having to work out general rules regarding a healthy diet. MyCookingCompanion incorporates several search functions that support the use of the database of recipes. Users can search by ingredients, combinations of ingredients, take into account preparation time, and their like or dislike of particular ingredients. Furthermore, the system supports functions such as creating a grocery list and keeping a record of recipes selected in the past. A key element in this version of the MyCookingCompanion is that recipes were classified as being more or less healthy, and that this classification was shown via easy-to-understand color coding. To this end, the healthiness of the 220 recipes in the database were rated on a 5-point scale (see figure 3.1) based on professionally endorsed nutritional facts such as overall calorie count and nutritional value. This classification was then verified by a dietician. Less-healthy recipes are indicated with a red or orange color code whereas healthy recipes are indicated with a light or dark green color code; each of the five categories contained approximately the same number of recipes. We decided to include not only healthy recipes in the database, but also less healthy ones for two reasons. First, including less-healthy but popular recipes that people might want to prepare occasionally would provide users with some choice. From a nutritional point of view this is not necessarily bad—eating less-healthy dishes once in a while is not itself unhealthy. And second, this would allow users to compare recipes and possibly lead to a better understanding of what constitutes a healthy meal.

We made a working prototype that can run on different hardware platforms, such as a laptop or a tablet, and used it to test the design of the concept. Figure 3.1 shows screen shots of our prototype.

Evaluation

We evaluated our prototype with potential users to collect feedback on such a concept. We also explored whether MyCookingCompanion would really help them make healthy changes to their main meal of the day. Forty Dutch women (23–58 years old) participated in this field study. All participants were living alone or with a partner and had no children living at home. Since we learned from many discussions with consumers that food habits change a great deal when children become part of a household, we decided that it would be too complex to include such a group in a first study. We asked the participants to discuss their participation with their partners first. All participants had expressed at the recruitment phase that they wanted to eat more healthy food and that for various reasons they had not always been successful doing this in the past.

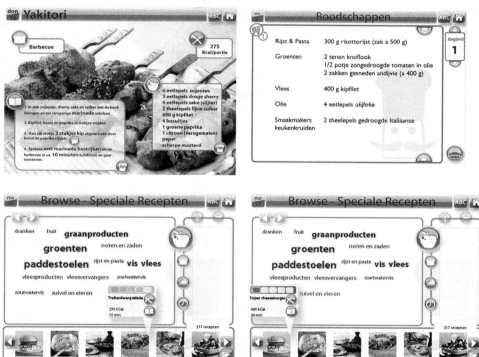

Figure 3.1

MyCookingCompanion selection of screen shots (in Dutch). Top screen shows the main page; middle left shows a recipe instruction page; middle right shows a grocery list; bottom left shows a recipe selection screen with the color coding indicating a more healthy recipe; bottom right shows a recipe selection screen with the color coding indicating a less healthy recipe.

We asked half of the participants to use a small laptop we loaned them with the digital prototype as described above; the other half we asked to use a paper version (like a cookbook), with the same set of recipes and the same layout on paper as the digital version. The participants were randomly assigned to the two conditions taking into account a balanced distribution over the two groups of age and household composition. We asked all participants to use the recipe source provided to them at home for two weeks, and in that period to select healthy meals, that is, meals coded green. We realized that providing them with this instruction would impact their behavior in a different way than if we had simply given them the system without such explicit instructions, but we decided that, given the setting of the study (with recruitment information and interview questions focusing on healthy eating), participants would be steered in this direction anyway.

We conducted three separate meetings with each participant to collect interview data: an intake interview, a brief check of first experiences after one week, and a closing interview. We recorded the semistructured interviews for later analysis, adopting content analysis techniques (Mayring 2000). In addition, we logged all use of the MyCookingCompanion. We also asked participants to keep a diary of the meals they had prepared in that period and take pictures of the prepared dishes, as an additional check.

Diary and Picture Record Results

Based on the diary information and the pictures, we could determine how many of their meals were healthy (i.e., coded green). In the laptop group, participants on average cooked 10.8 healthful meals in the fourteen days; this varied from five to fourteen healthy meals in the test period per participant. In the cookbook group, participants on average cooked 9.95 healthful meals; per participant, this varied from six to fourteen healthy meals. There was no statistically significant difference between the two groups.

Intake Interview Results

During the intake interview, participants were asked whether they had ever had the intention to eat healthy food but were not able to act upon that intention. Thirty-seven participants said that they frequently experience this situation. Too little time, laziness, social occasions,, and the appeal of unhealthy food were mentioned as the main reasons for that failure. In addition, several participants mentioned that healthy eating is not seen as "cozy"—a difficult to define label, but referring in this context to an atmosphere around dinner and eating. There were three participants who did not identify with this scenario and, instead, thought that they already ate very healthfully. We also asked all participants to recollect as accurately as possible the evening meals they had consumed in the previous two weeks, and then discussed with them to what extent these meals had been healthy or not. On average, they had had approximately 8.5 (more or less) healthy meals in the two-week period prior to the test. This represents a slightly lower average than during the test period. Of course, participants might have been inaccurate in their recollection, but this is something we could not verify.

We also discussed their knowledge about healthy food during the interviews. All participants mentioned public campaigns to promote consumption of fruit and vegetables, but used quite vague terms to describe them, such as "more fruit," "more vegetables," "less meat," "less fat," and "balanced." Twenty of the participants indicated that they were confident about their knowledge of healthy food, but that they do not always adhere to that knowledge. Twelve recognized that they possess limited knowledge about healthy food, and eight admitted that they do not know what healthy food entails. Table 3.1 presents an overview of the intake interview results, covering topics such as perceived advantages and disadvantages of healthy eating, what is keeping them from healthy eating, and what could support them.

Table 3.1
Overview of the intake interview results. The number in parentheses indicates the number of participants who mentioned this particular remark. Participants could provide more than one reason.

Reported advantages of healthy eating:
- Positive effects on short-term as well as long-term health (28)
- Healthy weight (25)
- Feeling fitter and more healthy (25)
- Body functions better (17)
- More energy (10)
- Positive effect on concentration (5)

Reported disadvantages of healthy eating:
- Takes more time to plan, shop, prepare, and cook (20)
- Less tasty or boring (16)
- More expensive (12)
- Not suitable for social events (such as eating with friends, birthdays) (7)

What keeps one from eating healthy:
- Lack of time (19)
- Eating at friends' or at restaurant, festive occasions (19)
- Desire to eat something tasty (which is seen as incompatible with healthy) (11)
- Laziness (11)
- Partner does not support healthy eating (9)
- Healthy food is not "cozy" (8)

What support to eat healthy would be welcome:
- Availability of healthy recipes on what and how to cook (20)
- More information on what is healthy (10)
- Where to store vegetables (10)
- Planning, discipline, time (each mentioned once)

Post-Test Interview Results

Most participants stated that during the experiment they ate tasty food, and that their knowledge about healthy food had increased. In general, participants liked that it was easy to choose a healthy recipe. They also explained that it was a good feeling to know that the recipe chosen was actually healthy and came from a trusted source; they knew that a dietician was involved in the choice of the recipes. Half of the participants mentioned increased confidence concerning healthy eating. However, two-thirds indicated that they would like more information on why something is unhealthy or healthy to enhance further the learning effect. In particular, participants mentioned that tips on how to change a recipe that is marked as unhealthy into a healthy recipe would be welcome.

In the post-test meeting participants were again asked to elaborate on the disadvantages of healthy eating and what might keep them from eating healthy food. The people that had mentioned before that taste could be a disadvantage had changed their mind on this aspect. Participants did not agree on whether "healthy eating takes more time": some really liked the recipes and thought that they were easy and fast to implement; others mentioned that planning, preparing, and cooking during the experiment took up more time than usual. A few participants mentioned at the third meeting that their expenditure on food had increased during the experiment.

All the participants mentioned busy professional and social lifestyles as factors that kept them from healthy eating. During the experiment they realized that it was more difficult to control their own food consumption and to stick to their intentions than they had expected at the beginning of the experiment. Examples mentioned were obligations at work, social obligations (e.g., meeting friends to eat out, having guests over) or the influence of their partner. This, according to participants, made it impossible for them to eat healthy food every day; thus, they would prefer then to be allowed to make choices on healthy eating in a broader time frame. For example, by having a weekly overview, that allows occasionally (e.g., once during that week) to select a meal that is not in the healthful category.

A total of twenty-seven participants also indicated a preference for a computer version of the tool. Their reasons included the following: it is easier to search with a computer; you can expand the database with your own recipes, and the program could calculate for you how healthy these are; you can make use of a social network to email recipes to friends, for instance; and you could search for recipes tailored to the ingredients you already have available at home. Eight participants preferred the book version because they were just not in favor of a computer for a task like this; they lacked sufficient computer knowledge, or they use a computer all day at work already. Five participants were indifferent about their choice and could mention advantages for both versions.

Thirty-seven participants either tried a new cooking technique, a new ingredient, or a new combination of ingredients and/or spices during the experiment. This is significant, since in our interview studies many participants indicated that they are often hesitant to try something new, because of the risks they perceive that it will not work out right, and the family will not appreciate the meal. Thirty-eight participants indicated that they had enjoyed being involved in the program and appreciated the new knowledge they acquired, the new recipes they tried, and being more aware of what they cooked and ate. Fourteen also indicated that even though they liked being "obliged" to cook healthfully, they also experienced it as a burden to cook regularly and to take the experiment into account when making appointments with friends or plans to eat out. All twenty participants allocated to the cookbook group wanted to keep the cookbook, a few even offered to pay for it. The twenty participants with the laptop version realized that they could not keep it (for copyright reasons it was made impossible to make a digital copy of the database); instead, several of them made written copies of recipes they liked.

The twenty participants who used the digital version of MyCookingCompanion answered additional questions concerning the platform, its functions, features, and interface. Preferences were quite diverse and personal; an overview of comments is provided in table 3.2.

We conducted a follow-up interview (on average four to five weeks after the experiment) with five participants of the cookbook group and five of the laptop group (recruited based on availability) to investigate effects on their eating habits in the longer term. All agreed that two weeks was too short a time in which to properly support them in habit change. They also agreed, however, that the experiment did help to create awareness about healthy eating. Furthermore, eight participants stated that they were still using the recipes they had copied from the tool, in order to maintain better habits.

Table 3.2
Feedback and suggestions provided by the participants using MyCookingCompanion

- Screen size—bigger might be better, to prevent having to scroll during cooking
- Touch screen instead of keyboard
- Able to put it up in the kitchen, but also take it elsewhere for relaxed browsing
- Appreciate functions such as: marking likes/dislikes of ingredients; grocery list; planning schedule; saving favorites
- Combine several search criteria at the same time, e.g., the time of cooking (15 minutes), the health indication (e.g., green) and ingredients
- Give suggestions on how to vary meals from day to day
- Suggest recipes taking into account the shelf-life of ingredients
- On demand information on why a recipe is healthy or not
- Tips on how to turn an unhealthy recipe into a healthy recipe
- Adding own recipes to the database, with the system providing a healthy indication and tips for improvements

Log-File Data

We recorded the usage pattern of all twenty participants using the digital MyCooking-Companion and analyzed the complete log file along with additional information provided by the participants in their diary. To ensure that the analysis was consistent among participants, we first defined what would be considered a browsing session (searching for recipes) and a cooking session (using a recipe description to prepare a dish). Rules were also established for identifying the start and end of a cooking and browsing session; for example, if a user left the system idle for longer than thirty minutes during a browsing session, we judged that the session had ended.

Usage patterns differed between participants ranging from 5 sessions and a total time of 1 hour and 56 minutes to 24 sessions and a total time of 16 hours spent using the system during the 14 days of the experiment. On average, participants had 14.8 sessions (7.5 hours) with 7.8 sessions (5 hours and 20 minutes) dedicated to cooking and 6.95 sessions (over 2 hours and 10 minutes) spent on browsing through recipes. The number of sessions is therefore almost equally distributed over cooking and browsing sessions; however, if inspected per time, participants spent 72 percent of the total time using the MyCookingCompanion to cook and 28 percent of the time browsing for recipes. The participants generally adhered to the instructions to use healthy recipes: most of the recipes viewed by participants were coded green. One participant decided to cook an unhealthy meal once during the experiment, two chose to cook two unhealthy meals, and one participant cooked three unhealthy meals. The participants who decided not to follow the test instructions explained that they, or their partners, liked these recipes.

Discussion

In terms of testing the sustainability of behavior change, two weeks is of course a relatively short period. However, given the practical problems involved in conducting longer-term field tests, it is not surprising that such studies are not often conducted. Despite the relatively short timeframe, the results of the research are valuable for understanding what helps and what hinders sustainable behavior change in meal patterns.

All participants indicated at the start of the study that they intended to change their diet. However, some of the participants learned during the test that they were not motivated enough to really change; instead, they preferred to continue with their cooking repertoire from before the study. Others were now even more motivated to continue on the path of changing their diets, indicating that the experience had helped them to get a better understanding of how to do that. Based on the analyses of the usage logs and interviews with the participants, we can conclud that a nutritional support tool (in either digital or paper form) can stimulate and increase a healthy food intake, but that success is definitely not certain if the participant is less concerned about healthy eating.

Many participants mentioned at the final meeting of the experiment that they had initially judged changing eating habits to be easier and more in their control than it actually turned out to be. Every participant experienced some strong barriers to having a healthy meal for fourteen consecutive days. It seems that in particular the social context in which food consumption decisions are made plays an important role, and that the participants did not take this into account when anticipating how successful they would be at compliance. Another strong barrier is linked to partners: if they are not supportive, it is going to be very hard to make changes. An interesting barrier identified by the research is the role of family conventions and traditions around regularly enjoying certain types of food; for example, in some households it is traditional to eat pizza every Friday or French fries during the weekend. The barriers experienced by participants during the study reflect several also mentioned in the literature (e.g., Branca et al. 2007; Mojet et al. 2001). However, the study also demonstrates that with easy-to-understand recipe guidance, people are able to adopt a more healthful food pattern, even if it is not every day. This last finding is also an important insight gained from the study: to provide optimal support, and keep people motivated, having a less healthy meal once in a while should be a formal part of any program.

A positive but "unofficial" finding of the study was that quite a large number of participants were hesitant to return the laptop or cookbook. They had learned to appreciate the support, the easy-to-follow coding, and the recipes so much that they found them hard to part with. And finally, food obviously is not just eaten for the nutrient values, but also for its taste. The experience made many participants realize that healthy food can also be tasty.

Conclusions and Future Work

Despite the importance of longitudinal research for understanding barriers in behavioral change, there has been few if any such studies reported in the literature. Our research contributes to addressing this gap. Of course, the significance of its findings would have been larger with a longer time period, a larger number of participants, and a more diverse set of household compositions involved.

The analysis of the usage logs and interviews with the participants suggests that a nutritional support tool (either a digital cookbook or paper version) can stimulate and increase a healthy food intake. But the results also highlight that participants need to have at least a basic drive to change their habits and eat in a healthier manner.

The analysis found no significant difference in results between the two test conditions—the digital cookbook and the hardcopy version—suggesting that the two formats, digital and print, do not seem to differ in their effectiveness. Since there was no condition in which the participants had to select recipes without the support of the color coding, it is not possible to claim with certainty that the color coding did make a difference. Participants did comment though that they appreciated the easy-to-understand color coding.

Our explicit instruction to prepare only green-coded healthy meals during the study was another factor in the results of this research. It would be interesting to see what happens if no such instruction were given. In follow-up studies, an additional condition with no support tool could also be added. Such a design would allow the study to test the effectiveness of the health indicator, that is, the color coding of recipes.

The interview results demonstrate that many participants did change their attitude toward healthy eating over time; participants indicated that they felt more confident about doing this compared to a situation in which they would have to make this decision on their own. Also, almost all participants changed their opinion about healthy food being "dull." Most participants enjoyed at least a subset of the healthy recipes and reported that they continued to cook these after their involvement in the research. These findings suggest the important role that knowledge development over time might play in truly sustainable habit change.

Acknowledgments

We thank all volunteers who participated in our tests and field study for their contribution. Special thanks to Jan Korst, Elke Daemen, Jan Engel, Marleen Twigt, and Marlies Wabeke for their support in this project.

DUTCH SNERT (PEA SOUP WITH BACON AND BLACK BREAD)

Makes 8–10 servings
For a traditional Dutch winter meal, serve the soup with black bread spread with mustard and bacon, and serve pancakes for dessert. To increase the healthiness of the soup, that is, to turn it from the slightly orange category to a green category, leave out the smoked sausage.

Ingredients

16 oz. split peas (dried)

2 quarts water

24 oz. uncured gammon (ham)

7 oz. uncooked spare ribs

1 smoked Dutch sausage (rookworst)

½ bunch celery

1 small celeriac

1 large carrot

2 leeks

Black bread

Mustard

Bacon

Preparation

1. Cook peas, water, gammon, and spare ribs approximately 1.5 hours till done. Meanwhile, wash and chop the celery and clean and dice the celeriac and carrot. Wash and slice the leeks.
2. Remove the meat from the pan and cube it.
3. Stir the peas to a mush and add meat, whole rookworst, celery, celeriac, and leeks. Dilute with water if necessary and simmer for 30 minutes. Season with salt and pepper.
4. Remove the sausage, slice it, and return the slices to the pan. The soup should now have porridge like consistency—snert is really thick soup.

Notes

1. See http://toolkit.s24.net/dietary-assessment.

2. See http://www.myfooddiary.com.

References

Ajzen, Icek, Cornelia Czasch, and Michael G. Flood. 2009. From intentions to behavior: Implementation intention, commitment, and conscientiousness. *Journal of Applied Social Psychology* 39 (6):1356–1372.

Baer, Adam. 2010. The most powerful food combinations. *Men's Health.* http://www.menshealth.com/mhlists/healthy-food-combinations/printer.php.

Barrier, Brooke. 2005. Nutrition expert: Consumer interest in health on rise. *Nation's Restaurant News* 39 (24):84.

Branca, Francesco, Haik Nikogosian, and Tim Lobstein. 2007. *The challenge of obesity in the WHO European region and the strategies for response.* Copenhagen, Denmark: WHO Regional Office for Europe.

Brug, Johannes, Anke Oenema, and Marci Campbell. 2003. Past, present, and future of computer-tailored nutrition education. *American Journal of Clinical Nutrition* 77 (4):1028S–1034S.

Chi, Pei Yu, Jen-Hao Chen, Hao-Hua Chu, and Jin-Ling Lo. 2008. Enabling calorie-aware cooking in a smart kitchen. *Lecture Notes in Computer Science* 5033:116–127.

Contento, Isobel R. 2007. *Nutrition education: linking research, theory, and practice.* Sudbury, MA: Jones & Bartlett Publishers.

Eurobarometer. 2006. *Health and Food.* Special Eurobarometer report no. 246. European Commission. http://ec.europa.eu/health/ph_publication/eb_food_en.pdf.

Geleijnse, Gijs, Peggy Nachtigall, Jettie Hoonhout, Liping Wang, and Li Qing. 2010. Promoting tasty meals to support healthful eating. In *Proceedings of the Wellness Informatics (WI) Workshop at CHI 2010.* New York: ACM.

Häubl, Gerald, and Kyle B. Murray. 2006. Double agents. *MIT Sloan Management Review* 47 (3):8–12.

Kraft, Pål, Filip Drozd, and Elin Olsen. 2008. Digital therapy: Addressing willpower as part of the cognitive-affective processing system in the service of habit change. *Lecture Notes in Computer Science* 5033:177–188.

Mayring, Philipp. 2000. Qualitative content analysis. *Forum Qualitative Sozial Forschung* 1 (2):20. http://www.qualitative-research.net/index.php/fqs/rt/printerFriendly/1089/2385.

Miller, W. C. 2001. Effective diet and exercise treatments for overweight and recommendations for interventions. *Sports Medicine* 31 (10):717–724.

Mojet, Jos, Elly Christ-Hazelhof, and Johannes Heidema. 2001. Taste perception with age: Generic of specific losses in threshold sensitivity to the five basic tastes? *Chemical Senses* 26:845–860.

Nestle, Marion. 2003. *Food politics: How the food industry influences nutrition and health.* Berkeley: University of California Press.

Paradis, Sebastien B., and Michel Cabanac. 2008. Dieting and food choice in grocery shopping. *Physiology & Behavior* 93 (4):1030–1032.

Riebe, Deborah, Bryan Blissmer, Geoffrey Greene, Marjorie Caldwell, Laurie Ruggiero, and Kira M. Stillwell. 2004. Long-term maintenance of exercise and healthy eating behaviors in overweight adults. *Preventive Medicine* 40 (6):769–778.

Rosin, Odelia. 2008. The economic causes of obesity: a survey. *Journal of Economic Surveys* 22 (4):617–647.

Walls, Helen L., Anna Peelers, Bebe Loff, and Bradley R. Crammond. 2009. Why education and choice won't solve the obesity problem. *American Journal of Public Health* 99 (4):590–592.

World Health Organization. 2011. Fact sheet: Obesity and overweight. Fact sheet No. 311, updated March 2011. http://www.who.int/mediacentre/factsheets/fs311/en.

4 Not Sharing Sushi: Exploring Social Presence and Connectedness at the Telematic Dinner Party

Robert Comber, Pollie Barden, Nick Bryan-Kinns, and Patrick Olivier

We increasingly use technology to connect with other people, and the growing diversity of our interactions is mediated through and with technology. As shown by this book, one such interaction is our sharing of food experiences. Looking to overcome the growing distances (physical and temporal) within which relationships are built and maintained, individuals and groups are turning to computer-mediated communication (CMC) to share food with those they love. This trend of sharing mediated food experiences, and the recent move toward models that promote the celebratory aspects of food practices in the field of human-computer interaction (HCI) (Grimes and Harper 2008), raises a number of important questions for the study of the interactions among people, food, and technology: How do we virtually share something which is as innately physical and sensorial as food? How do we share the social nature of a meal, when we cannot share the physical food on which it is based? How do we mediate the performance of celebration around shared food? How do we express ourselves (and our connections to others) through shared food when we cannot share physical food or spaces with those others?

With these questions and others we explore what has become a significant interest in HCI research on ubiquitous, tangible, and embodied interaction: the mundane orientations, interactions, and configurations of individuals, artifacts, and practices in shared space (Dourish 2001). We extend this research by exploring the situated and timely unfolding of physical and social phenomena at the Telematic Dinner Party (TDP), a technology system for mediating remote guests in a shared dinner party experience.

Telematic Dinner Party

With the telematic dinner party (TDP) we explore the mediation of food, food practices, and the shared connections people build around food. While there has been much work done in the arena of supporting intimate and familial relationships (Gaver 2002; Goodman and Misilim 2003; Kaye and Goulding 2004; Vetere et al. 2005) the dinner party provides a hybrid space for public and intimate interactions, where existing relationships can be strengthened and new relationships formed (Williams 1996). With the TDP, the research aim goes beyond sharing past dining experiences (e.g., Urban Spoon, Yelp, Foodspotting) to sharing an embodied dining experience among a group when not co-located. The TDP also explores the hybrid space of real and "not" real or inverse presence (Timmins and Lombard 2005). Inverse presence is where people describe "real" experiences as mediated ones (e.g., "it was like a movie"). Here we are mashing up the real and mediated to explore how this can support an overall shared experience.

Related Work

The Chit-chat Club (Karahalios and Dobson 2005) employs a physical avatar to share a table in a public café with a local group. Reinterpreting face-to-face communication for a distant guest, the remote person is alone and "dials in" to join a co-presence group. The asymmetry of the one-to-many interaction puts the burden on the remote participant to remain engaged with the present group, and for the present group to reconfigure itself to the artifact and person that is the avatar. The project pushes the boundaries of mediated social presence, but fails to recognize the problematic nature of re-embodiment on the social dynamics at the dinner table.

Attempting to facilitate the remote sharing of food, CoDine (Wei et al. 2011) translates the messages of love and respect that are expressed in the ritual of most Asian cultures of serving food to loved ones. Rather than attempting to recreate the nuances of serving, CoDine employs a 3D printer to literally write messages of love and respect in edible goo. Thus, despite the difference in how food is shared and served, the meaning behind the gesture is maintained with the recipient being able to consume an offering of love and respect. Yet, the sense of agency produced across the space must be reconfigured to attend to the printer and the printing process. While traditional dinner practices involve passing food to others, the 3D printer involves a more deliberative process. Thus, attending to the interface between diners may remove the diners from the dining experience.

Remote agency is also observed as problematic in the NetPot project (Foley-Fisher et al. 2010), which investigates communal cooking as a platform for computer-mediated connectedness. The NetPot provides a central platform where both local and remote participants "cook together" over the same pot. Here, much like CoDine, the cooking experience is asymmetrical: the local guests have access to the real food, with taste, smell, and texture, while the remote guests are "cooking" food icons and are represented as avatars. The end result is that while remote and local cooks may perform the act of cooking together and commune around the NetPot, the asymmetries in practice also reflect asymmetries in the meaning of the shared experience. The actions of the local guests have real consequences for what they might consume, but also for what the remote guests can interact with. Thus the physical and unshared food is the platform for engagement and the product of only one set of guests' engagement.

These projects all have the technology explicitly exposed in the experience. There is no mistaking the avatar for a person, food icons as edible, or the mechanical printing of foodstuffs foreign to a dining experience. For TDP, we seek to retain the dining experience where the interactions and focus remain on the table and fellow guests. The design and implementation of the platform intentionally acknowledges that TDP is not a fully co-present experience but provides the salient aspects of the experience: interactions, offering food, seeing the actions of eating, and localized presence at the table, for example.

The Design

The design of the TDP system was supported by observations of traditional dinner parties and exploratory studies with a low fidelity prototype environment (see figure 4.1 and for an overview of the iterative development of the TDP, see Barden et al. 2012). Framed as an issue of social presence (Biocca and Harms 2002), we identified practices of shared dining that might serve as points of departure in considering the design of the TDP. We explored the sense of shared social presence (Biocca and Harms 2002) that emerges through those practices, creating a sense of shared space, shared engagement, and shared (inter)activity.

From the early studies we identified embodied practices of sharing food (either physically or through presenting food visually to remote guests), toasting as a means to signify communion, the sharing of experiences around and through artifacts, and the dynamics of communication at the table including side conversations and the frequent dissolution of conversation between dislocated groups (Barden et al. 2012).

Figure 4.1
Low-fidelity prototype environment for the telematic dinner party. The local and remote guests are seated and projected at opposite ends of the table. Toasting (left) and messaging (right) require guests to attend to separate dining spaces.

From this, and the HCI literature, we identified three interrelated challenges for the design of a system for the mediation of shared dining: (1) designing for shared space and re-embodiment of remote actors at the dinner table, (2) designing for remote agency around food, and (3) designing with and for the social structures and dynamics of the dinner party.

Shared Space and Re-embodiment

In the observational studies of dinner parties, all guests shared a physical space. This shared space supported communion and celebration by affording numerous practices, such as guests offering each other food. With the exploratory prototype system, the aim was to replicate this dining experience, but in the absence of the shared physical location. Due to the distance between locations (London and Barcelona) the system was set up using commercial video conferencing and projection equipment available at both locations. Owing to the relatively poor quality of the equipment it was necessary for the projection and recording to occur in separate spaces (see figure 4.2). Unlike the observations at dinner parties, the guests in the prototype system now had to contend not only with being physically dislocated, but also in the sense of their own location, being virtually dislocated. With the design of the TDP we actively seek to overcome this second form of dislocation.

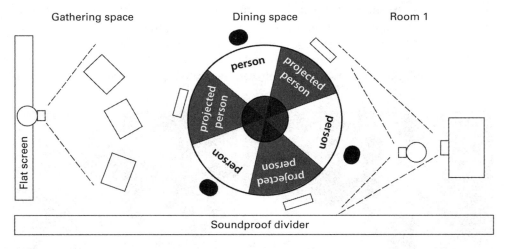

Figure 4.2
The setup for one location in the telematic dinner party. Three guests sit at a round table, with projections of remote guests, captured and relayed from overhead, interspersed between them. Seats are positioned at each of the six guests' position, with those for remote guests having only audio equipment in them. At the edge of the dining area a screen and camera relay a wide view of the space, intended to be used at the beginning of the meal.

In the TDP, two physically dislocated groups of three guests are seated at round tables (see figure 4.2). The intention with the TDP is to translate and support, to some extent, the shared space of a traditional dinner table. At each table three individuals are seated 120 degrees apart with a place setting at each of their own seats. Using an overhead HD camera and projector, the actions of the individuals are captured and relayed to the other table. The actions of the dislocated individuals are interspersed between the physically present guests, projected from above. The projections occupy the space between guests, filling the dinner table. The top-down perspective is chosen to integrate all the guests at the table and allow shared cues such as the actions of eating and playing with food. This integration of remote guests also addresses the asymmetry observed in the Chit-chat Club (Karahalios and Dobson 2005).

Each guest also wears a lavalier microphone that is connected through a soundboard to a speaker co-located with their projection at the dislocated dining table. Unlike the prototype system, which used a single speaker and commercial video conferencing software, the TDP audio system provides for high-fidelity directional perception of presence. Although the speaker cannot direct their speech, their speech is perceived to originate with their projection by the dislocated dinner party guests. Directional audio perception was added to this iteration of the TDP in order to overcome issues of distinguishing dislocated guests from each other and to enhance the sense of embodiment of the dislocated individual in the physical space. In effect, this creates a sound environment that translates and supports in the TDP platform the occurrence of side conversations observed in the original co-located dinner parties. This also moves further from the use of physical and virtual avatars to represent embodiment to reflect the rich communication and sensory environment of a traditional dinner party.

Remote Agency

Sharing food was a central practice in the observed dinner parties that was critically lacking in the prototype system (Nowak and Biocca 2003). With the TDP a deliberate design decision was made to support a literal interpretation of this shared interactivity through a set of network turntables (lazy Susans). The lazy Susan is a traditional dinner table artifact, which is placed at the center of the table. Food is placed on the lazy Susan, and it can be rotated by guests to get food for themselves or to share actual food with others. Including such a means for interactivity followed from observations at both the co-located dinner party, where guests shared food and artifacts at the table, and while using the prototype system, where guests used the video projections to share items, interact with each others' food, and so on.

In the TDP, the lazy Susan turntables were intended to allow for the actions of one group to be relayed to the other group and vice versa. The turntables were given the simple task of coordinating their movements: when one turntable is turned, the other responds and moves to the same specified angle. This system was designed to allow individuals the opportunity to offer each other food: by spinning one turntable the food on the other (observed by the projection) could be presented to a dislocated other. The turntables also extended the sense of shared space as the projection of items placed on the turntable was mirrored in the other room. It further served as a means for individuals to contact each other, to draw attention from the remote room and to playfully tease the others by moving food away from them (Nowak and Biocca 2003). In one telematic dinner party, two dislocated guests took turns to interrupt each other as they reached for food. In others, the guests placed their food in identical positions on the turntables so they both "ate" from the same dish. The literal interpretation of the sharing of food with the lazy Susan turntables attempts to restore the embodiment of food sharing, often lost in the digitization of food experiences.

Social Structures of the Dinner Party

Much like a traditional dinner party, the TDP is designed to involve collaborative tasks in the form of various party games. These are intended to merge the groups in endeavors that can be supported by or promote shared engagement. These activities are chosen by the co-hosts of the dinner party and included a murder mystery party, telematic Pictionary, a show-and-tell of items that remind individuals of their home country, and sharing songs and stories from the cultures of guests. These activities themselves do not offer shared engagement (a guest or group can choose not to engage in them), but provide a platform from which shared engagement might be built. In this sense, shared engagement is a feeling of shared experience and collaboration (Nowak and Biocca 2003). Further shared engagement could also be expected from the act of dining itself, where individuals share experiences with and of the food, and construct their shared experience of the TDP, and dinner parties more broadly.

The TDP also incorporates a "gathering space," a space to the side of the main dinner party area that allows for a soft start to the meal. When guests arrive they can interact with each other through a set of cameras and monitors, which referenced a more familiar interaction. These tools are intended to be peripheral to the main dinner party and only be used at the beginning of the party and if there are any technical difficulties with the table-based system. In one of the first sessions, this peripheral technology became a distraction for one of the hosts, who constantly monitored it, trying to get a better view on the reactions of the remotely located guests. In subsequent dinners, this peripheral view was turned off as the table-based system was technologically stable enough to support the interaction. This peripheral system draws attention to the vulnerability of the translation of the dining experience to competition from the accepted or expected modes of interaction.

Telematic Dining

Having designed the TDP to respond to challenges of mediating shared dining practices, we sought to explore the changing meaning of social presence and connectedness
in the TDP environment. Four telematics dinner parties were run with the setup
described above. Each party had a unique theme and activity selected by two hosts.
Each host also invited two guests. Two dinner parties involved international exchange
students visiting the United Kingdom from China, and the remaining two dinner
parties involved guests from or living in Newcastle. Meals were observed and feeds
from all video and audio feeds were recorded. Participants were interviewed at the end
of each dinner party. The video feeds (top and long view) from the TDP, were used to
analyze the guests interactions during the dinner party. We were able to compare their
reporting in the post-dinner interviews, with their actions during the dinners. In order
to address the questions raised in the introduction, we explored in this data the ways
in which guests appropriated or avoided the TDP platform in their performance of the
celebration of the dinner party. In the success and failures of the TDP to create a sense
of shared social presence, we identify three topics for the future design of mediated
environments for shared food experiences: food as a medium for presence, the reinterpretation of the embodiment of guests and their actions, and the support and
structures for the social dynamics around the table.

Food as a Medium for Presence

While the guests at the TDPs reported achieving some sense of shared space, engagement, and interactivity, there was one meal where this connection was not as successful. Our attention is brought to this meal by the comment of one guest:

I don't feel that I shared food with them. It felt like we [were] together in one room and they
were eating in another room. There was no sense that we were sharing. (guest J, telematic Pictionary dinner)

Food is a physical object that has a number of qualities that we (so far) cannot
transmit digitally. Any number of these qualities can denote a shared physical space,
such as the smell, warmth, or feel of food. When we are in a shared dining experience,
we eat and taste and experience food simultaneously. We can use that to relate our
experiences of the food, to be engaged in the same activity, and to be sharing the social
and visceral experience of eating. In the telematic dinner party, food is not shared in
this way. At best, and as it was for the dinner parties, food is mirrored across spaces.
Thus, eating together has a different meaning.

The guests at all but the telematic Pictionary dinner party quoted above were able
to share food experiences. Although guests could not directly share the food they were

eating, they were provided (through the projection and networked turntables) with a means to observe the dislocated others eating and interact with them and their food, to engage in the act of sharing food through a literal mediation. However, in the telematic Pictionary dinner party the sense of shared space provided by the mirrored projection of food was obscured. The sushi platter that the guests ordered was both large enough and sufficiently dark to obscure the projection of the remote platter. Only when one platter was removed could the food consumed by the remote guests be seen (see figure 4.3). This occurred only twice in the meal since it was cumbersome and an interruption to lift the platter. When the platter was lifted the guests were surprised at the differences in what, and how much, either group had consumed.

Here we note that both the shared space of the turntable and the shared engagement of experiencing sushi together (one guest in each room was invited as they had never tried sushi before) are lost. Without these the interactivity of the turntables also becomes meaningless. In fact, it was experienced only as interference. Unlike other

Figure 4.3
At the telematic Pictionary dinner party, the guests' sushi platter filled the networked turntables and obscured the projection of the remote platter. Only when one was removed could the guests observe the remote locations food. Here one group has consumed far more of their platter than the other.

dinners where individuals could see what others were eating, offer them new foods, or even tease them by interrupting them, the absence of the shared engagement in eating meant, in the telematic Pictionary meal, that the movement of the turntable was experienced as a sign of disengagement. In contrast to the other TDPs, the movement of the turntables was not coordinated or intentionally playful and served only to interrupt people when they reached for food. Both groups were unaware of the actions of the other, creating layers and asymmetries in interaction. Thus responding to the notion of food as a medium also brings to the fore the re-embodiment of the individuals as they respond to the presence or absence of the physical and sensorial aspects of food.

Reinterpreting Embodiment at the Table

The dinner table is a site for the communion of family and friends. In contrast to eating on the go, or in front of the TV, it symbolizes shared space as a necessity for eating together. However, as geographical distances between friends and families grow and time constraints limit opportunities for commensality, there is a need to re-interpret the notion of shared space at the dinner table. When individuals do not share a physical space, we must attempt to create a sense of that shared space.

In general, the TDP system was perceived by participants to create a strong sense of shared presence. The projection was perceived as useful and as a point of play. All participants expressed some sense of novelty and engagement with the projections at the early stages of the dinner parties. Such novelty was expressed through actions such as hand holding, attempting to high five, and playing with each other's food. Placing a projected hand over a remote other's was both a means to express closeness but also used as a means to create playful discomfort. Similarly, playing with remote guests' food was seen as a means to playfully interfere in the other's dining experience, enhancing the sense of togetherness.

However, like the Chit-chat Club, the projections could not convey the full range of expected interactions and guests perceived limitations on two points: (1) the low quality of the image and (2) the overhead position hindered direct facial emotional communication between physical and remote guests. The low fidelity of the video presence was illustrated by the physical guests' lack of respect for the dining space of the projected others. In many cases the remote guests' projections and place settings were covered with discarded plates, napkins, and food. This lack of recognition of physical presence at the table occurred regardless of whether a physical place setting was maintained or not for the remote guests. In contrast, the directional audio achieved a sense of presence such that, in one dinner party, when a guest dropped a fork on the floor, guests in the remote location turned to see where the fork had fallen. At other points, guests would talk directly to the audio equipment that filled the chair of a remote guest.

Although the system is intended to represent individual actions as integrated with those of others, there remained asymmetries of action and perception for guests. The guest from the telematic Pictionary party quoted earlier also experienced much of the TDP as operating on different "layers" of interaction. Specifically, he experienced a sense of his physically co-located guests as "up here" (at eye level) and the dislocated guests as "down there" (on the projected tabletop). The primacy of the physically co-located also drove individuals to disconnect from the remote guests when engaged with the co-located others. In the telematic Pictionary dinner, the CMC afforded by the TDP remained as a backdrop to the physically co-located, in the participant's words, as a "screen saver." The physical embodiment of otherness and the response to that physicality was critical for the experience of disconnectedness. This contrasted to other dinner parties, where the projected images were seen as a source of playfulness, becoming shadows and shadow puppets. Thus, it is not inherently the projections, but how they are appropriated, that includes them in the dining experience.

This also points to the limitation experienced by participants in the lack of emotional expression in the top-down projection. This reiterates the claim of Jaron Lanier (cited in Mantovani and Riva 1999) that the manipulation of the perspective of the individual impacts on the meaning of the communication. Here the withdrawal of the facial expression of emotion was a limitation on communication. Yet, this is not to say that the TDP was devoid of emotion, but rather that guests were required to translate their emotions to new forms, such as playful teasing, hand holding, and high fiving projections.

Supporting Social Dynamics

The use of existing conventions for party games laid the basis for successful engagement across the telematics dinners. In all but one party, guests expressed a sense of shared engagement, and admitted feeling connected to their remote guests through their activities. When guests shared narratives and songs of their home countries or showed items that had personal meaning for them, others could relate to these and also share their own experiences. In the murder mystery party, the guests also worked closely together to construct a shared narrative. In all three of these dinners, the guests also actively engaged with each other's food. Guests shared food, played with each other's food, and talked about the food they were eating. With the overhead projection, they were able to easily see the remote guests eating, and find ways to interact with the food.

For the telematic Pictionary, where the sense of connection and engagement was not strongly felt, the game was reserved for after the dinner and separate from the meal itself. The theme for the dinner was that the guests were requested to wear "elegant red and black attire." The guests reported that the directive to dress up was a nice way to separate the event from a casual meal. However, they did not find the theme to have much meaning during the dinner itself:

I did like that we had to do something to participate a bit more. That increased curiosity. But the theme itself wasn't that relevant during the dinner so it wasn't prominent then at all. I like that everybody was dressed up made it a bit more serious. Something special. During the dinner it wasn't that relevant to me at all. (guest A, telematic Pictionary dinner)

The fact that they were dressed in similar colors and styles was not mentioned during the dinner, nor was the theme carried into the meal itself, such as "red and black" food. The other three dinners utilized elements of a theme to encourage the guests to engage and prompted conversation among the guests. The strictly aesthetic and cursory nature of the "red and black" theme did not provide the common ground or shared talking points that the other more integrated themes supported.

Summary

The telematic dinner party is an exploration of the intricate social practices of celebratory shared dining in a mediated environment. We began with questions of how we might share food and participate in the social, celebratory, and performative features of a shared dining experience through a computer-mediated environment. These issues rest upon supporting a shared connection, or a sense of "togetherness" for guests at the telematic dinner party. We note that although in general guests experienced some sense of shared presence and connection with one another, the limitations of the sense of shared celebration experienced in the telematic Pictionary dinner party point to asymmetries in the sharing of food experiences across mediated environments. The platform of a shared dining experience brings together these issues of space, agency, and interaction in the food itself, where food is a medium for presence, an artifact around which agency occurs, and a platform for social engagement. In the absence of the shared food experience, the shared dining experience falls apart. Not sharing sushi meant not sharing in the experiences of eating sushi and not sharing togetherness.

The meaning of food as an artifact and space, however, is also reflexively designated by the shared space, agency, and interaction that it affords (Dourish 2001). While food is central to the dining experience, food also acts as an artifact and space around which individuals orientate themselves and their interactions. While the TDP provides mechanisms for individuals to share space through re-embodiment, this space can easily be obscured by artifacts and by the actions of others. Despite having place settings and meals projected into remote locations, virtual guests' spaces were intruded upon by the waste of physical guests. In the destruction of this shared space the value of a shared meal is also reduced. Without this space, the remote guests can no longer share a meal and their actions become invisible or meaningless. On either side of the projections guests separately consume the same meal. In this case, not sharing the visual and physical properties of sushi meant not sharing the actions and interactions with that sushi.

Such actions in the dining space take on meaning depending on the dynamic and constructed social context. The act of dining has its own etiquette regarding interaction with other guests and with food. In the Western tradition, you are expected to ask for an item to be passed to you rather than reaching across the table, for example. The networked lazy Susans were designed to translate this "passing" ability in the mediated space. In this final case, not sharing sushi was a transgression from expected modes of dining, of participation in the dining experience. This created frustrations and misinterpretations of the movement of the lazy Susans. While not successful in the telematic Pictionary party, in the other three dinner parties they were used as a point of play, particularly in attempts to subvert access to the food or as a means to get attention. While this teasing counters the tradition of etiquette it is not outside the realm of levity that would be expressed in a traditional dinner party. As a consequence, play and performance become significant considerations in the TDP (see Barden et al. 2012). The fostering of these two elements demonstrates the importance of having a platform open enough to allow the participants to infuse their own methods and meanings in the system. This parallels the expectations for a dinner party, which has rules but is also flexible and allows one to knowingly "break the rules" and expand the space of dining to include levity and play (Salen and Zimmerman 2003). Thus, the social structures and dynamics of a shared dining experience are not only critical to the facilitation of remote connectedness, but are open to re-interpretation as a source of connecting experiences.

Conclusion

We designed and implemented the telematic dinner party environment to critically examine the limits of technological mediation of social presence, celebration, and communion at the dinner party. The dinner party typically offers a space to explore new and maintain old relationships through shared practices, social structures, and the sharing of food and experiences. In the TDP, the qualities of shared space, engagement, and interactivity are explored through re-embodiment, remote agency, and social dynamics. The findings point to a number of asymmetries of experience that highlight the interwoven nature of engagement, activity and space, and the fabric of social practices which can be supported at the dining table. In connecting through food, individuals at the TDP are invited to share food, share party games, and to share connected, yet dislocated, spaces. However, the relative success or failure of the TDP is spread across a number of social and technological factors where the guests perform celebration and communion. We suggest, following Gunawardena (1995), that the focus of technology to mediate communication must reflect social and not technological concerns. This provides a requirement for individuals to have some sense of shared interaction based on remote agency and through shared space. Moreover, we bring to the fore the necessity to consider food as a medium and space around which individuals configure themselves and their actions and interactions. We suggest that the embodied connection to others through food is established not in the food itself alone (e.g., CoDine), nor in the communication that occurs at the table (e.g., Chit-chat Club), but in the coming together of actions and interactions around and with food.

Acknowledgments

This research was part funded by the EPSRC SiDE project and Marie Curie Action under the European 7th Framework Program Balance@Home project. This work is supported by an EPSRC Doctoral Training Centre EP/G03723X/1 (HE), an EPSRC Leadership Fellowship EP/G007144/1 (MDP), and EPSRC IDyOM2 EP/H013059/1. Additional support was provided by Furtherfield and Latitudinal Cuisine in London, United Kingdom, and Telenoika in Barcelona, Spain.

ROB'S "YOU CAN'T MAKE FRIENDS WITH SALAD" CHEESECAKE BROWNIES

Makes 1 to many servings, depending on appetite and available diners
I'm not much of a cook, so I have only a few fall-back recipes for those dinner party moments. This one is a relatively simple brownie recipe, but with an added touch of cheesecake to keep them moist, and if you get a nice swirl on top they look great. Unfortunately, brownies might not be the healthiest contribution to the meal, but then you can't make friends with salad. Serve with ice cream or fruit, or just cut into bite size chunks and leave within arm's reach.

Ingredients

Metric measurements included here for cooks who like to weigh their ingredients for exact proportions.

4 oz. (125 g.) butter

4 oz. (125 g.) dark chocolate, chopped

8 oz. (200 g.) brown sugar

3 eggs, lightly beaten

2 oz. (50 g.) flour, sifted

2 oz. (50 g.) cocoa powder

¼ tsp. baking powder

8 oz. (225 g.) cream cheese, softened

2 oz. (50 g.) caster sugar (or baker's sugar, which is more finely granulated than table sugar sold in the United States, but not as fine as powdered sugar; it is available in supermarkets or online through Amazon.com, or it can be ground at home in a food processor)

Preparation

1. Combine butter, chocolate, and brown sugar in a medium saucepan and stir over a medium heat until chocolate and butter have melted. Remove from heat and whisk in eggs.
2. Add flour, cocoa, and baking powder and stir until well combined. Pour into a greased and lined 8 × 11 inch rectangular baking pan.
3. Beat cream cheese and caster sugar until smooth and creamy. Spoon it randomly over chocolate mixture and swirl with the tip of a knife.
4. Bake at 300°F for 35–40 minutes or until cooked through. Allow to cool before slicing.

References

Barden, Pollie, Rob Comber, David Green, Daniel Jackson, Cassim Ladha, Tom Bartindale, Nick Bryan-Kinns, Tony Stockman, and Patrick Olivier. 2012. Telematic dinner party: Designing for togetherness through play and performance. In *Proceedings of ACM Conference on Designing Interactive Systems 2012 (DIS2012)*, 38–47. New York: ACM.

Biocca, Frank, and Chad Harms. 2002. Defining and measuring social presence: Contribution to the networked minds theory and measure. In *Proceedings of the 5th International Workshop on Presence*, 7–36. http://ispr.info/presence-conferences/previous-conferences/presence-2002/

Counihan, Carole, and Penny Van Esterik. *Food and culture: A reader*. New York: Routledge.

Dourish, Paul. 2001. *Where the action is: The foundations of embodied interaction*. Cambridge: MIT Press.

Foley-Fisher, Zoltan, Vincent Tsao, Johny Wang, and Sid Fels. 2010. NetPot: Easy meal enjoyment for distant diners. *Entertainment Computing* 6243:446–448.

Gaver, Bill. 2002. Provocative awareness. *Computer Supported Cooperative Work* 11:475–493.

Goodman, Elizabeth, and Marion Misilim. The Sensing Bed. In *Proceedings of UbiComp 2003 workshop*. New York: ACM.

Grimes, Andrea, and Richard Harper. 2008. Celebratory technology: New directions for food research in HCI. In *Proceedings of the 26th Annual SIGCHI Conference on Human Factors in Computing Systems 2008 (CHI'08)*, 467–476. New York: ACM.

Gunawardena, Charlotte. 1995. Social presence theory and implications for interaction and collaborative learning in computer conferences. *International Journal of Educational Telecommunications* 1 (2):147–166.

Karahalios, Karrie, and Kelly Dobson. 2005. Chit-chat club: Bridging virtual and physical space for social interaction. In *Proceedings of CHI'05, extended abstracts on human factors in computing systems*, 1957–1960. New York: ACM.

Kaye, Joseph, and Liz Goulding. 2004. Intimate objects. In *Proceedings of the 5th conference on designing interactive systems: processes, practices, methods, and techniques (DIS '04)*, 341–344. New York: ACM.

Mantovani, Giuseppe, and Giuseppe Riva. 1999. "Real" presence: How different ontologies generate different criteria for presence, telepresence, and virtual presence. *Presence (Cambridge, Mass.)* 8 (5):540–550.

Nowak, Kristine, and Frank Biocca. 2003. The effect of the agency and anthropomorphism on users' sense of telepresence, copresence, and social presence in virtual environments. *Presence (Cambridge, Mass.)* 12 (5):481–494.

Salen, Katie, and Eric Zimmerman. 2003. *Rules of play*. Cambridge: MIT Press.

Timmins, Lydia, and Matthew Lombard. 2005. When "real" seems mediated: Inverse presence. *Presence (Cambridge, Mass.)* 14 (4):492–500.

Vetere, Frank, Martin Gibbs, Jesper Kjeldskov, Steve Howard, Florian Mueller, Sonia Pedell, Karen Mecoles, and Marcus Bunyan. 2005. Mediating intimacy: Designing technologies to support strong-tie relationships. In *Proceedings of the SIGCHI Conference on Human Factors in Computing Systems (CHI '05)*, 471–480. New York: ACM.

Wei, Jun, Xuan Wang, Roshan Peiris, Yongsoon Choi, Xavier Roman Martinez, Remi Tache, Jeffrey Koh, Veronica Halupka, and Adrian Cheok. 2011. CoDine: An interactive multi-sensory system for remote dining. In *Proceedings of the 13th International Conference on Ubiquitous Computing (UbiComp '11)*, 21–30. New York: ACM.

Williams, S. 1996. *Savory suppers and fashionable feasts: Dining in Victorian America*. Knoxville: University of Tennessee Press.

5 Civic Intelligence and the Making of Sustainable Food Culture(s)

Justin Smith and Douglas Schuler

There is a growing movement of people seeking to establish a more sustainable, and culturally relevant agri-food system than the current industrial-style food system. For some, this is about reclaiming food traditions that evolved over the course of history and are nearly lost. For others, it is about creating entirely new food cultures situated in particular places where producers and eaters are "working together" to address the ecological, social and economic impacts of a global industrial food system (Lyson 2004; Wright and Middendorf 2008; Friedland 2008). Thomas Lyson coined the term "civic agriculture" as a way to conceptualize a range of emergent alternatives to the mainstream food system (2004). These alternatives emphasize community and ecological diversity, proximity, scale-appropriateness, and civic engagement.

For example, in the Pacific Northwest, the indigenous peoples of the Coast Salish are collaborating to reclaim a food culture rooted in both the land, and in their oral histories (Swinomish Tribe). Further inland, farmers and food activists are also working to protect the fertile Skagit Valley by re-creating a vibrant local food economy that is feeding families while providing small-scale farmers a living wage. Just to the south, in Seattle, neighborhood associations are developing sustainability programs centered on the practice of urban agriculture. And east of the Cascade Mountains, in the Palouse region dominated by commodity wheat production, university students, farmers, and an engaged citizenry are collaborating around the values of a diversified food system that address the needs of the community. These include providing fresh fruits and vegetables to low-income families through "backyard gleaning" programs (Backyard Harvest), as well as establishing community gardens for people who don't have access to garden space. The spread of civic agriculture is also evident across Washington State through the enactment of public policies that are putting Washington-grown foods in public school cafeterias (Fresh Food in Schools). This reflects the collective effort of citizens, media savvy advocacy groups, as well as state and local political leaders. It also highlights the evolution of public choice and the recognition of the importance of teaching children about the importance of how our food is grown and where it comes from.

All of these examples represent what Douglas Schuler (2001; 2008) refers to as civic intelligence. One can identify people working together, sharing their values, ideas, and expertise, as well as financial and technical resources, in order to address perceived social problems in their communities. Civic intelligence most commonly refers to the ways organizations network, identify constraints, and facilitate opportunities in order to address serious problems of a civic nature (Schuler 2001; 2008). These include climate change, poverty and hunger, nuclear proliferation, and many other social ills. Aside from engaging in political advocacy many groups are developing practical solutions, coalesced around the idea and practice of how we produce and share the foods we eat within our geographic communities.

But civic intelligence is about more than just the ways people and organizations share resources for civic purposes. The concept includes the ways in which a networked civil society learns over time, adapts to changing circumstances, and improves its practice. To help understand the various aspects of the concept, we provide a visual description of Schuler's model (figure 5.1).

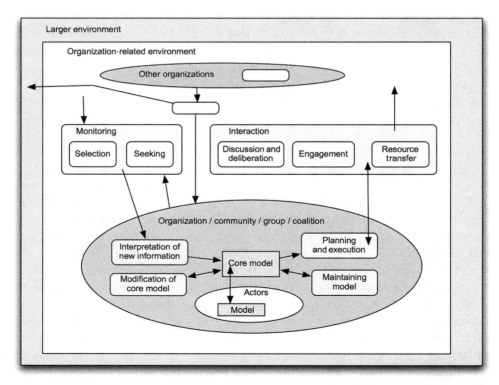

Figure 5.1
Civic intelligence model.

The model presents a conceptual framework for those looking to understand and improve the capacity of an intelligent and responsive civil society. Civic intelligence focuses on the "organization as agent," and the ways organizations "monitor" their environment, access and process information, and ultimately coordinate with other organizations. Schuler's model provides a holistic conceptualization of the ways that ordinary citizens and civil society collaborate to solve shared problems.

Objective

This chapter is concerned with understanding an active civic intelligence and its place in shaping sustainable agri-food systems. The scope of our study is specific to the Pacific Northwest of the United States. Within this region we limit the scope further by focusing on Washington State. Washington represents a unique case in terms of ecological and sociocultural variations. The eco-regions exhibit distinct differences in terms of climate, topography, biotic formations, and human settlement patterns (see figure 5.2).

As Raymond Jussaume and Kazumi Kondoh (2008) note, these differences have played an important role in shaping the face of agriculture in Washington State (similar

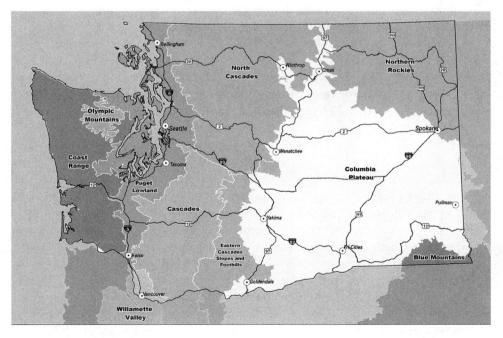

Figure 5.2
Washington State eco-regions.

in impact to the political context of each region). Furthermore, when linking the ecological with the social and political histories of each sub-region, it becomes apparent that this larger context is central in determining the constraints and opportunities that exist for making a sustainable food culture and, by extension, new agri-food systems.

While consideration of these factors is important, we are driven by a shared interest in the praxis of social change. We seek to strengthen civic action by identifying those patterns of communication and action that are proving effective. By focusing on the influence of place and civic intelligence, we hope to highlight reusable patterns that are relevant to the histories and cultures of the people who inhabit the state.

Building on research conducted at Washington State University in 2011, we draw insights from results derived from the analysis of network structures between organizations promoting urban agriculture across the major metropolitan areas of Washington State in the Pacific Northwest of the United States. Considering both the content and programs that organizations develop as examples of their ideas in action, we assess the knowledge and ideological themes that are informing the character of food movements across the Pacific Northwest. In addition to improving our understanding of civic intelligence, we are motivated by an interest in improving our civic intelligence in terms of solving common problems in our communities.

The transition toward sustainable agriculture and, more specifically, sustainable food cultures will require citizens to do more in creating the change they want to see. At the same time, effective civic action will require greater capacities to perceive problems, as well as to communicate and coordinate in direct contradiction to dominant social forces. Similarly, effective civic action centered on a reshaping of our agri-food systems will also be sensitive to the local particulars of place, including the history and culture, as well the social and ecological aspects of distinct geographies.

Context

Distinct food cultures have emerged over time, born out of the complex interplay between human ingenuity and specific landscapes (Grigg 1995). Just as the environments of particular places and regions have shaped, enabled, and constrained human civilizations—and, thus, the capacity of individuals and societies to learn, share new ideas, and trade their culinary treasures—this unfolding has reshaped the very character of those environments that influenced those first civilizations. In some cases, our effectiveness at transforming the physical environment has resulted in the collapse of our communities and even entire civilizations (Diamond 2005). With respect to the ways that modern society is currently producing food, there is a growing concern about the social and ecological consequences of the industrial food system (Pollan 2006; Wright and Middendorf 2008).

In today's societies, the ability to reshape the physical landscape is unprecedented. The communication structures that enable societies to organize and coordinate around the production and exchange of goods and services are without parallel, both in terms of their complexity and pervasiveness. Manuel Castells marks this latest human revolution as the "Information Revolution" and the "Rise of the Network Society" (1998). A central feature of this transformation has been the further de-linking of human activity from the constraints of space and time in the traditional sense. Of particular importance has been the rapid growth of global corporate power enabled through the adoption and diffusion of information and communication technologies.

With respect to agriculture, food products can be ordered and paid for from anywhere, capital moved from one country to another, all with a click of a button from a computer with an Internet connection. GPS-operated farm equipment is also aiding in the advance of industrial scale agriculture, notifying farmers where, and how much to harvest based upon the needs of their contractors (Lyson 2004; Lyson and Guptill 2004; Sadler et al. 2005; Wolf 1998). These and other technical systems are responsible for the management and logistics of a global industry of cultivation, processing, and distribution (Lyson and Guptill 2004). New information and communication technologies (ICTs) provide tools for increasing the diffusion of conventional agricultural practices by opening up new channels for information access among farmers. Although old diffusion patterns persist, there is a growing use of ICTs, which are often a requisite for farmers seeking to survive in the industrial food system (Wolf 1998).

This process of diffusion also includes a particular form of discourse that perpetuates further industrialization and modernization of our agri-food systems as the only means to feed the world in a sustainable manner. This is connected to the growth and consolidation of power in the global commodity food chain by large agribusiness. Such consolidation of power constrains farmers, giving a handful of multinational corporations control of nearly every aspect of the production, processing, and distribution of food commodities (Lyson 2004). As a result, farmers are compelled to work as employees of these global entities because there is no alternative (Pollan 2006). Similarly, consumers are largely dependent upon these same corporate actors in the provisioning of food (Boucher 1999; Lyson 2004). The foods that consumers have access to are the result of the economic rationales of large scale-food production and the actors that market these products.

The underlying structuring and rationalization of our agri-food systems have advanced a particular food culture, especially within the Global North. Consumers residing in the richer nations often expect to find oranges in January; at the same time (Pollan 2006), consumers also expect to find foods that are convenient to prepare, neatly packaged, and consistent in quality and appearance. Yet, the economic rationality used in meeting these expectations (at a global scale) require an incredible amount of precision and coordination that have only been made possible through ICTs and the technological advances in cultivation, processing, storage, and distribution (Wolf 1998; Lyson and Guptill 2004).

Just as technology can provide advantages vis-à-vis the coordination and consolidation of political and economic power in our agri-food industry, citizens, civil society, and activists are using these same technologies to organize, communicate, and assemble in opposition to the growth of this globally controlled form of large-scale agribusiness. From local nonprofits to international NGOs, people are contesting the logic and impacts brought about through the industrialization and globalization of our agri-food systems (Wright and Middendorf 2008). ICTs are enabling organizations to articulate alternative solutions to mainstream systems of production and exchange. Food activists are using these tools to share information and organize around community food projects, such as food security programs, community kitchens, and edible urban landscapes. We find an information-rich movement, networked through ICTs, enabling the diffusion of shared ideas such as cultivation practices, as well as prescriptions for a "new moral economy of food" (Morgan et al. 2006).

With the diffusion of these practices, local ecologies of place are shaping the ways in which communities and civil society are implementing and experimenting with these alternative food paradigms. These interactions with new ideas and specific constraints of place are shaping the character of emergent food cultures. By considering the structure and character of civic intelligence expressed by those working to create a new food culture, we seek to identify common patterns that work to strengthen the capacity of change agents to bring about new ways in which we produce and exchange food.

Review of Research Methods

Websites were the primary data used for analysis. This suggests some initial assumptions about the usefulness of web data for understanding the relationship between civic intelligence and sustainable food cultures in the region. The first of these assumptions is that hyperlinks provide easily accessible data on the relationships that exists between organizations (Lusher and Ackland 2011). These links provide qualitative data when viewing the website as text/discourse. Hyperlinks can point to organizations that provide financial and human resources, whereas other links might refer to ideological or knowledge resources (videos, scholarly articles, news, etc.) that help in coordinating the collective work of these interrelated organizations (Smith and Jussaume 2010).

Mapping hyperlink structure can reveal the types of themes and projects that these groups promote following a simple text analysis. We can also determine to some extent the ways in which these organizations do or do not collaborate on projects, whether these projects represent urban community gardens, fresh food delivery to seniors and low-income peoples, or elementary-level food- and farming-education programs. The use of website data can also help to determine both the physical location of the organization as well as the physical location in which organizations carry out their work.

Finally, we make the assumption that food cultures are rooted in place, requiring, at the very least, an understanding of the features that constitute these socioecological landscapes. By bringing together a social network perspective to the study of civil society, along with an analysis of textual and spatial aspects of these efforts, we can paint a more complete picture of the ways in which communities, cities, and organizations are working together to construct sustainable environments that feed and increase the well-being of people.

Following the research conducted at WSU, the organizations sampled for this study were drawn from a series of queries using the Google search engine. In our query string we combined, "Washington State," "Western Washington State," "Eastern Washington State," "Puget Sound," "Palouse," and "Columbia Valley" with phrases such as "sustainable food," "local food movements," "urban food movements," "urban agriculture," "sustainable food culture," and "local food culture."

We chose five organizations operating within the state. We focused on search ranking, the organization's location, and operation scale (e.g., neighborhood, town or city, county, regional or state level). The initial websites included: UrbanFarmHub, The Washington Sustainable Food and Farming Network, the Washington State Farmer's Market Association, Tilth Producers of Washington, and Sustainable Solutions. While the majority of organizations are specifically focused on food and agriculture, others take a broader focus on sustainability and development.

Findings

The linkages between organizations revealed a great deal of information about the emergence of civic intelligence and the making of sustainable food cultures in the Pacific Northwest. The analysis showed a dense, interrelated set of organizational connections between groups that focus on and operate within the state. There were also additional sets of weaker linkages to national level groups. One of the interesting things we found was that there was a high degree of connectivity among urban-centered food groups and regional organizations. Yet, a number of these linkages were highly centralized around various government agencies from Seattle and King County public health departments, and the Washington State Department of Agriculture (WSDA). We found significant linkages between all groups and the state university and extension system through the University of Washington and Washington State University. Network structures are altered when considering the nature of the relationships between groups, however, whether these relationships are primarily financial or information/knowledge-centric in nature.

Many of the linkages showed a high degree of financial relationships, but centered on only a handful of organizations. But we should note that these might not be wholly indicative of the nature of fiscal interactions between groups. It is just as probable that individual fiscal interactions with organizations are as important. Furthermore, it is possible that these individual financial backers are also affiliated with other related groups working on similar issues. Unfortunately, we lack individual response data for organizational members. This left our interpretations limited to information the groups have chosen to share through their websites.

It is interesting, though, that despite the centrality of links with government and financial support, the informational linkages are much more distributed, lacking a highly centralized structure. This suggests that there may be a high level of information exchange and coordination among groups around ideology and framing, but it might be dependent on the types of information connected. In the context of new agricultural research in the area of organic and sustainable food production we find centralization of knowledge exiting with universities and national-level agricultural organizations. While all groups are information producers in some respects, shared information is tightly focused on a small handful of network participants.

In the case of neighborhood- or city-level organizations operating within urban and suburban landscapes, we find more distinct networks that exist almost independently of one another, except where those neighborhoods are linked through shared resources such as finance, information, and project collaboration. Again, we find some linkage to government and university support. Even with government support, however, many organizations are filling information gaps not being addressed by either government or scholarly research at the major universities in the region.

Civic organizations, governments, and academia coordinate information by sharing and transferring it through a range of information and media types. The mode of transfer is often linked to the intended audience. For instance, we found a number of garden groups using video to showcase successes, as well as for advocating concepts of sustainability, economic development, and happiness. Ideological interactions are generally horizontal in nature when organizations have opportunities to access and share information or reproduce it in alternate frameworks. Only in the context of new agricultural knowledge and financial resources do we see a vertical network structure, generally with state- and national-level organizations coordinating and disseminating knowledge of best practices. The same was found with those information links between government and universities. When reviewing the network structures that emerge with respect to the sharing of scientific knowledge and financial resources, only a handful of organizational actors operate as central channels for the larger set of city, county, state, and regional organizations. At the same time, the sharing and reproduction of movement ideology is shared across most organizations, aside from the larger scientific and financial intuitions supporting new research in the area of agriculture and food systems.

It shouldn't be assumed that government agencies and universities hold a monopoly on the production of new knowledge or of the financial resources that are coordinated in the creation and dissemination of food system information. In fact, a number of garden organizations and nonprofit urban farming organizations actively engage in forms of citizen science that are being used to help educate new gardeners and urban farmers. Yet this remains only a small subset of the agricultural information being disseminated. For instance, many of these organizations act as conduits, working to synthesize and simplify complex scientific research into practical tools for a nonacademic movement of young farmers operating across both rural and urban landscapes.

Landscapes and Place

In terms of spatial differentiation, we found that urban environments possessed the greatest level of networked interactions, as well as the greatest level of political momentum toward transformation of the industrial food system. This was evident regardless of the regional differences among all nine eco-regions in Washington State. This makes sense when considering the centrality of urban centers to the function of the regions they serve. Each eco-region possesses one or more urban space that functions as a center for the coordination and emergence of action around creating new food cultures.

The Puget Sound Lowlands operates as the cultural and social center of western Washington. In this area there is a high degree of spatial heterogeneity across the landscape, which consists of urban areas, forests, and diversified agricultural lands (based on Landsat data). Most of the agricultural spaces are predominately made up of small to midscale farms that produce a range of products and services (USDA 2007). With their proximity to urban centers across the region, these farms have been able to respond to the shift in consumer preference toward more reflexive values consistent with a vibrant civic agriculture.

Historically, where the urban spaces remain small or isolated, such as Wenatchee or Spokane, the growth and sustainability of these types of farms have been less apparent. Evidence exists, however, of substantial shifts occurring in these cities as organizations and local government seek to promote increased public health through new policies that support farmers as well as consumers of all socioeconomic backgrounds (Spokane County Dept. of Public Health 2010). Yet the slowest changes appear to be occurring in those landscapes which remain isolated or that have even smaller urban centers. This is apparent in the Columbia Plateau eco-region where commodity agriculture for international markets continues to dominate the agri-food landscape.

Through the growth of support networks and organizations such as the Tilth Producers of Washington and the Washington Sustainable Food and Farming Network, however, more and more producers in this region are adopting more sustainable production practices while diversifying their marketing efforts to target a growing consumer base interested in local and region foods. This is being supported by an increasing number of restaurants in the various urban centers that are proactively experimenting with producing cuisine that is of a uniquely northwest character. Again, these changes appear to be accelerating as more people and organizations share ideas, influence each other, and support a shared vision for promoting a sustainable food system in Washington State and throughout the Pacific Northwest.

Conclusions: Patterns Identified

We have chosen to organize these findings as patterns following the format laid out in *Liberating Voices* (Schuler 2008), which was derived from Christopher Alexander's model of patterns and pattern languages (Alexander et al. 1977). Alexander's use of the word "pattern" refers to a distinct element of knowledge that describes a design problem in the field of urban planning and architectural design.

Each pattern describes a problem that occurs over and over in our environment, and then describes the core of the solution to that problem in such a way that you can use the solution a million times over without ever doing so the same way twice (Alexander et al. 1977).

A pattern language is a network of patterns that call upon one another. Patterns help us remember insights and knowledge about design and can be used in combination to create solutions (Alexander et al. 1977).

Since Alexander's first articulation of patterns and pattern languages, the approach to structuring "reusable" knowledge has been adopted by software engineers (including the influential "Gang of Four" book on design patterns, Gamma et al. 1994), regional planners (Van der Ryn and Cowan 1996), as well as scholar-activists working to support democratic forms of communication and problem solving (Schuler 2008). Although we are not describing an entire pattern language around the making of sustainable food cultures, we do offer a rather extensive list of patterns that reflect the range of practices identified through our research.

The patterns that we identified in our research were well represented in the *Liberating Voices* pattern language. In listing a set of identified patterns we attempted to present them in order from most important (table 5.1). We defined the most important in terms of their foundational role, and their occurrence regardless of the social and ecological context in which they occur. While we feel that all of the patterns represent important features in the creation of a sustainable food culture in the Pacific Northwest, some do appear to be more or less invariant in positive social change.

Table 5.1
Associated patterns by ranking

1. Civic Intelligence	11. Techno-Criticism
2. Education and Values	12. Earth's Vital Signs
3. Shared Vision	13. Open Action and Research Network
4. The Good Life	14. Public Agenda
5. Opportunity Spaces	15. Informal Learning Groups
6. Strategic Capacity	16. Citizen Science
7. Strategic Frame	17. Intermediate Technologies
8. Political Settings	18. Community Animators
9. Whole Cost	19. Media Literacy
10. Community Networks	20. Mobile Intelligences

The most obvious pattern that exists again and again is of course "civic intelligence." Intelligence is the product of both biological and social evolution and in this sense it historically grounded; it is imperfect and still evolving (Dewey 1939; Schuler 2001). The emergence of these social networks is the outcome of adaptive processes and the creativity of individual and collective intelligences as a means to develop new patterns of living and working together in human society. These patterns are undergirded by aspects of shared values and the co-creation of a shared vision to ameliorate the social and ecological problems that face our communities. This vision of the "good life," of what that is, and what it is not helps orient purposeful action in a positive direction.

SQUASH BLOSSOM "TAMALES"

Makes 4 servings, perfect when piled next to a bowl of refried black beans with a sprinkle of cojita or goat cheese on top.

The following recipe was posted on the Neighborhood Farmer's Market Alliance website. The Neighborhood Farmers Market Alliance (NFMA) is a community-based, nonprofit organization that operates seven farmer/food-only markets in Seattle neighborhoods. The NFMA's mission is to support and strengthen Washington's small farms and farming families. The organization does this by providing effective direct sales sites for the region's small farmers, and by educating consumers about local farm products and the importance of supporting and preserving local farmland.

Jamaica Jones, who provided the recipe, has yet to meet a vegetable she doesn't like. Jamaica is working as a freelance writer and researcher, while she is pursuing a master's degree in food studies from New York University. Upon graduation, she hopes to write a book chronicling the footprint of Arabic food traditions around the world.

Squash blossoms—typically the male blooms of both summer and winter squashes—prove a wonderful addition to countless dishes, offering sunny color, a fine sweet flavor, and a long-standing culinary legacy across the Americas. When choosing blossoms, be sure to select those that are not yet dried or withered at the top. Before cooking, rinse them lightly with cool water to remove any hitchhiking bugs. The more you experiment with squash blossoms, the more you'll discover how versatile they are. Irresistible when simply battered, fried, and sprinkled with salt, squash blossoms can also hold their own when stirred in, at the last minute, to risottos and quick pastas. They marry happily with hard cheeses like Asiago and bear a particular affinity for their seasonal counterparts: corn, tomatoes, and basil.

Ingredients

8 fresh squash blossoms, stamen end removed

¾ cup yellow cornmeal

½ cup all-purpose flour

1 tsp. baking powder

½ tsp. salt

½ tsp. freshly ground pepper

¼ cup grated Asiago cheese (or any comparable farmers market cheese)

3 tbsp. fresh basil, chopped

2 farm-fresh eggs

¼ cup milk

1 tbsp. melted butter

Salt and pepper to taste

2 tbsp. oil for pan frying

¼ cup chicken stock or water

Preparation

1. Prepare the blossoms: gently open each one to remove the stamen and pistol, and discard them. Twist the blossom from the squash. Hold the blossoms in your hand under running water and rinse, then set them aside to dry on paper towels.

2. Sift 1/2 cup of the cornmeal and the flour, baking powder, and salt into a medium mixing bowl and toss in the pepper, Asiago, and basil. In a smaller bowl, combine 1 egg, the milk, and the butter.

3. Add the wet mixture to the dry, stirring just enough to combine all ingredients.

4. Using a narrow spoon, coax 1–2 tbsp. batter into each blossom, tuck in the bottom of each blossom, and gently twist the top to close (a fresh chive can be used to tie the blossom shut as well). Set the blossoms aside.

5. In a small bowl, lightly beat the remaining egg. To another bowl, add the remaining 1/4 cup cornmeal, and season with salt and pepper.

6. Heat the oil in a heavy skillet over medium heat. Dredge each blossom first in egg and then in the seasoned cornmeal and add them, as space permits, to the skillet. Do not overcrowd them in the pan, working in batches if necessary; turn them as needed to brown all sides. This should take about 5–6 minutes. Turn the heat to low, add a tablespoon or two of broth, cover, and allow the blossoms to steam for another 5–6 minutes. Serve immediately.

References

Alexander, Christopher. 1979. *A timeless way of building*. New York: Oxford University Press.

Alexander, Christopher, Sara Ishikawa, Murray Silverstein, Max Jacobson, Ingrid Fiksdahl-King, and Shlomo Angel. 1977. *A pattern language*. New York: Oxford University Press.

Altieri, Miguel A., and Peter Rosset. 1999. Ten reasons why biotechnology will not ensure food security, protect the environment and reduce poverty in the developing world. *AgBioForum* 2 (3&4):155–162.

Boucher, Douglas H, ed. 1999. *The paradox of plenty: Hunger in a bountiful world*. Oakland, CA: Food First Books.

Connell, David J., John Smithers, and Alun Joseph. 2008. Farmers' markets and the good food value chain: a preliminary study. *Local Environment* 13:169–185.

Diamond, Jared. 2005. *Collapse: How societies choose to fail or succeed*. New York: Penguin.

Dewey, John. 1939. *Intelligence in the modern world: John Dewey's philosophy (edited and with an introduction by Joseph Ratner)*. New York: Modern Library.

Friedland, William H. 2008. Agency and the agri-food system. In *The Fight Over Food: Producers, consumers, and activists challenge the global food system*, ed. Wynne Wright and Gerad Middendorf, 45–67. University Park: Penn State Press.

Gamma, Erich, Richard Helm, Ralph Johnson, and John Vlissides. 1994. *Design patterns: Elements of reusable object-oriented software*. Reading, MA: Addison-Wesley.

Grigg, David B. 1995. *An introduction to agricultural geography*. London: Routledge.

Jarosz, Lucy. 2000. Understanding agri-food networks as social relations. *Agriculture and Human Values* 17:279–283.

Jussaume, Raymond, and Kazumi Kondoh. 2008. Possibilities for Revitalizing Local Agriculture: Evidence from Four Counties in Washington State. In *The fight over food: Producers, consumers, and activists challenge the global food system*, ed. Wynne Wright and Gerad Middendorf, 225–246. University Park: Penn State Press.

Lyson, Thomas A. 2004. *Civic agriculture: Reconnecting farm, food, and community*. Medford, MA: Tufts Univ. Press.

Lyson, Thomas A, and Amy Guptill. 2004. Commodity agriculture, civic agriculture and the future of US farming. *Rural Sociology* 69:370–385.

Lusher, Dean, and Robert Ackland. 2011. *A relational hyperlink analysis of an online social movement. Journal of Social Structure 12 (5).Morgan, Kevin, Terry Marsden, and Johathan Murdoch. 2006. Worlds of Food: Place, Power and Provenance in the Food Chain*. New York: Oxford University Press.

Norton, George W. 2009. *The economics of agricultural development: World food systems and Resource Use*. New York: Routledge.

Pollan, Michael. 2006. *The omnivore's dilemma: A natural history of four meals*. New York: Penguin.

Sadler, Edward J., Robert G. Evans, Kenneth C. Stone, and Carl R. Camp. 2005. Opportunities for conservation with precision irrigation. *Journal of Soil and Water Conservation* 60:371–378.

Schuler, Douglas. 2001. Cultivating society's civic intelligence: Patterns for a new "world brain." *Information Communication and Society* 4 (2):157–181.

Schuler, Douglas. 2008. *Liberating voices: A pattern language for communication revolution*. Cambridge: MIT Press.

Smith, Justin, and Raymond Jussaume. 2010. Rhetoric and realities of social equity with respect to agri-food systems. Paper presented at XVII ISA World Congress of Sociology, Gothenburg, Sweden, July 11–17.

Spokane County Department of Health. Personal Communication. June 15, 2010.

United States Department of Agriculture (USDA). 2007. Census of Agriculture 2007. National Agricultural Statistics Service. Accessed at: http://www.agcensus.usda.gov/Publications/2007/Full_Report/usv1.pdf

Van der Ryn, Sim, and Stuart Cowan. 1996. *Ecological Design*. Washington, DC: Island Press.

Wolf, Steven. 1998. *Privatization of information and agricultural industrialization*. Boca Raton, FL: CRC Press.

Wright, Wynne, and Gerad Middendorf, eds. 2008. *The fight over food: Producers, consumers, and activists challenge the global food system*. University Park: Penn State Press.

COOK

Yvonne Rogers and Kenton O'Hara

The advent of pop-up restaurants—appearing in trucks roaming the streets, shacks in run-down parts of downtowns, and other outdoor urban spaces—has contributed to our rethinking about how food is cooked, purchased, served, and enjoyed. Hipsters tweet and "thumbs up" their latest discoveries and tourists follow. Not only are customers enjoying a new culinary experience but chefs are also relishing the new challenges it brings. Typically, chefs are much more constrained by the space, equipment, and facilities available to them compared with traditional tethered cafés and restaurants. But rather than work around these, they work with them, proudly mastering and showing off their skills in making original and delightful meals with very little at hand. A central theme is serving up traditional, modern fusion, or other favorite dishes using locally sourced ingredients. Some chefs also play with the space and place in which they work. Ben Spalding, for example, a rising star in the United Kingdom's foodosphere, spent a summer in a school playground in the East End of London at a makeshift stall where a maximum of eight customers sat around him on simple chairs. He had no grill, gas, or fridge, just some pre-cooked food from the day before and a stack of Tupperware containers of raw ingredients that he mixed and melded in front of their noses. Foodies and locals flocked to his stall to literally dine out. They were not disappointed. Playing on the idea of eating outdoors, he even dished up his fruit dessert in a garden trowel (figure II.1).

Besides the global movement toward new creative cooking located in unusual places, more and more people are rethinking their own cooking, eating, and food-shopping habits. The role of technology in these endeavors is an emerging area of research, as is a focus into the relationship between obesity and healthy eating practices. Until now, many approaches to obesity have been at an informational level where appropriate dietary knowledge and prescriptive programs are presented with the expectation that people who are overweight will follow them. The contributing role of behavior in weight gain, as the *root* of poor eating patterns, has often been overlooked. In chapter 6, Esther Toet, Bernt Meerbeek, and Jettie Hoonhout take such behavioral factors as the source of their design enquiry into *solutions* for poor eating

Figure II.1
Ben Spalding's fruit dessert served up in a garden trowel at his Stripped Back pop-up restaurant.

patterns. They focus in particular on the notion of "mindless" eating and the various ways that we can over-consume without thinking. Drawing on the social scientist B. J. Fogg's model of behavior change, and with a view to making eating more mindful, they engage in a user-centered design process to devise and test a series of conceptual prototypes for food preparation and eating. The concepts take a range of existing kitchen utensils, crockery, and cutlery, and augment them with various sensing, timing, and notification technologies. This augmentation is designed to address the behavioral concerns of mindless eating, by helping people choose smaller portions, adopt slower eating patterns, and make more conscious decisions about additional helpings of food.

Concerns with behavioral change are also at the heart of Grant Young and Penny Hagen's work discussed in chapter 7. Here, though, the focus of the research lies not at the point of preparation and eating, but at the point of purchase. More specifically, they investigate how to support behavioral change around the consumption of locally sourced, fresh, seasonal produce. As with Toet and colleagues, one of their motivations for researching the affects of behavioral change relates to improved health outcomes. But for Young and Hagen, there are a broader set of motivations at play in the choice to consume local and seasonal produce, including a reduction in the impacts of transport, increased food security, improved flavor, and local socioeconomic benefits. In light of these motivations, the chapter introduces the FlavourCrusader, a social mobile seasonal produce guide designed to facilitate local and seasonal consumption choices. Their design responds to identified barriers to local consumption practices such as limited access to farmers' markets and lack of knowledge about produce seasonality. Such information is not enough in itself, however. The app is also designed to support the meal planning process—the broader context for such information—through the provision of a searchable recipe database accessible right at the point of purchase. While the authors highlight the potential for such guides, they also raise some of the more pragmatic concerns with their development and maintenance. Here, their discussion points to the potential of social networks and related platforms that can be used to leverage existing data sources and communities in the ongoing development of these seasonal consumption resources.

The theme of local food sourcing continues in chapter 8. Eric P. S. Baumer and co-authors offer further insight into the cultures and practices surrounding farmers' markets. The complexity of motivations and attitudes shaping practices in these settings makes them rich sites for design enquiry but also presents some methodological challenges. The chapter offers a methodological way forward in the form of cultural probes, and discusses the authors' experiences deploying these in a farmers' market in New York, with a view to understanding how to design for these settings. The particular aim here was to move from the more formal and functional concerns of traditional "requirement gathering" to one that encapsulated the more subjective attitudes toward the aesthetic experiences of the market place.

Chapter 9 takes an alternative perspective on the local sourcing of food. In contrast to the purchase of locally sourced food, Katharine Willis and co-authors consider the practice of "scrumping": the foraging of fruit growing in the wild. Here the emphasis is not so much on reconstructing a relationship between producers and consumers, but on how, through picking apples and pears from trees on roadsides, people re-embed their relationship with food into a meaningful spatial structure of their lives. The authors explore this through the notion of "mundraub," which is the practice of picking fruit from trees in public spaces. More specifically, they present a case study of mudraub.org, a website intended to support fruit foraging by viewing or tagging public fruit tree locations on an open access interactive map. Aside from highlighting the practical benefits of facilitating fruit sourcing, the website shows how its the users begin to develop a deeper understanding of the seasonal nature of the fruits. Moreover, through engagement with the practice, people became more aware of the places they live in and began to see urban places, such as car parks and roadsides, differently. Through the sharing of place-based information, a rich sense of communal co-presence is developed. Mundraub is not as motivated by economics as it is by a moral compulsion to harvest what has been given, to not see it go to waste but accept what the trees are offering as a gift. From such moral underpinnings to these activities arise important social aspects where individual gain gives way to the harvesting and sharing of surplus fruit.

So it is not just chefs who are transforming the culinary experience of cooking and dining. The chapters in this section show how researchers have also been developing Web, mobile, pervasive, and social networking technologies to inform and nudge people to change the ways they approach their eating and cooking practices. Just as a piece of chicken can look and taste quite different when served on a brick so, too, can wild apples taste and look quite different when discovered and plucked from a roadside tree. The trick is to keep finding joy, surprise, and pleasure in preparing, eating, and understanding food. The chapters here give a glimpse into the many ways technology can help us do that.

6 Supporting Mindful Eating with the InBalance Chopping Board

Esther Toet, Bernt Meerbeek, and Jettie Hoonhout

In our affluent Western society, food is easy to obtain. High-calorie food is presented to us everywhere and is also apparently hard to resist: obesity is a growing problem, with an increase in a range of health issues attributed to being overweight, such as cardiovascular diseases and diabetes. In the past, excess weight and obesity were considered problems that could be simply solved by adhering to a diet prescribed by medical practitioners or dieticians. But recent studies indicate that weight gain and obesity is to a large extent a behavioral problem, that is, most overweight and obese people experience problems with controlling or regulating their eating behavior (Jansen et al. 2009; Van Buren and Sinton 2009). The strong behavioral aspect of obesity and weight gain explains the poor long-term results of most diets. It is estimated that less than 5 percent of those who have lost weight after dieting will have maintained these losses after four to five years (Kramer et al. 1989). In the long run, a diet does not change the behavior patterns that caused the weight gain. It is not sufficient to know how to lose weight: the psychological processes that lead to behaviors associated with weight gain should also be dealt with (Jansen et al. 2009).

One of the reasons why changing eating behavior can be difficult is that many of the decisions concerning food intake are made, as Wansink calls it, "mindlessly" (Wansink 2006, 276). In exploratory interviews on mindful and mindless eating, we learned that many people who actively try to manage their weight are already aware of the (mindless) calories consumed during snacking. At dinner time, however, they often forget that too large a portion will mean that they consume too many calories, even if the ingredients are considered healthy. It seems as if they lose control when they finally have the feeling they are allowed to eat. Roy F. Baumeister and colleagues (1998) compare the amount of self-control of an individual with a rechargeable battery: if the battery drains there is no self-control available until the battery is recharged, and that takes time. Following the reasoning of these researchers, we can state that when dieters restrain themselves from eating snacks, there is no self-control left during dinner, and they consume more than intended. The interview participants indicated that they consider the consumption of these unintended calories to be problematic, as the behavior occurs daily and therefore has a big impact on their total food intake. In addition to this lack of self-control, a lack of knowledge about healthy portion size and the influence of external cues (e.g., size of packaging or a plate) make it more difficult to choose or assess healthy portions (Wansink 2004; Jansen et al. 2009; Van Buren and Sinton 2009).

Obviously, there are many aspects to a person's lifestyle or behavior that can contribute to the problem of obesity. Likewise, there will be many ways to address this problem on a societal, group, and individual level. A combination of these approaches will be required to reduce the problem of obesity, but in this study we specifically focus on mindless eating.

Following a user-centered design approach, we investigated how to make people more mindful about their eating behavior during dinnertime in order to persuade them to adopt a healthier lifestyle. To establish long-term behavior change, we aimed for a product that fits with regular cooking habits, and matches the user's cooking skills and habits. To further limit the scope of our design space, we decided to focus on end-consumer solutions that can be used during the cooking process, make use of state-of-the-art technology, and that furthermore do not require a fundamental change in other parts of the current food value chain (e.g., in agriculture or food production and distribution).

Design Directions

We explored three design directions that we thought might influence the consumption of excessive, mindless calories during dinner:

1. Dish up appropriate portion sizes.
2. Slow down the eating rate.
3. Wait before taking second servings.

The first direction, regarding appropriate portion sizes, focused on a fact mentioned in the introduction, that many find it difficult to choose their portion size. They have little knowledge about what constitutes a healthy portion and are influenced by external cues, including the size of their plate or packaging. The second and third design directions focus on creating awareness of one's satiety level. The physiological mechanisms for satiety are rather complex and still being unraveled (Woods, 2004). But a distinction can be made between long-term signals induced by the hormones leptin and insulin and short-term satiety signals induced by the ghrelin hormone. Although the hormonal system informs us that we are satiated, these signals and the corresponding feeling of satiety are not always perceived in time. This may cause people to overeat.

We analyzed the three design directions with the behavioral model for persuasive design of B. J. Fogg (2009). In order to be persuaded into a particular target behavior

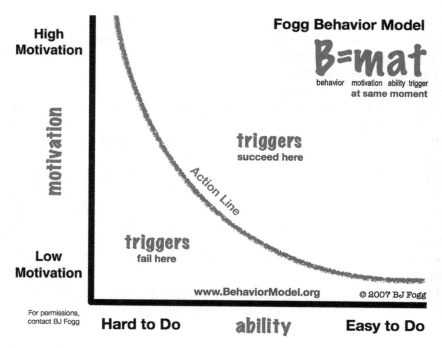

Figure 6.1
Schematic representation of the behavior model for persuasive design (Fogg 2009).

(e.g., eating with a slow pace) a person should, according to Fogg's model, have both the motivation and the ability to perform that behavior. Moreover, there should be a trigger that activates the user to perform the target behavior. Fogg explains that an effective trigger for a small behavior can lead people to perform harder behaviors. Since many dieting attempts end up in failure, it could be assumed that losing weight and weight management is difficult and is an example, as such, of a "hard" behavior. Therefore, the ultimate target behavior is not necessarily the same as the target behavior of the design, but in the end should lead to it.

Figure 6.1 schematically represents the behavioral model. On the x-axis is the ability of people to perform a particular target behavior and on the y-axis is their motivation to perform that behavior. A target behavior is very likely to be performed if people are highly motivated and have the ability to perform it. The core motivators for human behavior, according to this model, are pleasure/plain (related to sensation), hope/fear (related to anticipation), and social acceptance/rejection (related to social cohesion). For example, the "pleasure" that is a motivator for mindful eating could be the enhanced taste experience when people eat more slowly. To increase people's ability to perform the target behavior, Fogg refers to simplicity factors as being: time, money, physical effort, brain cycles (cognitive effort), social deviance, and non-routine. A simplicity factor for eating mindfully could be to make information about the healthiness of ingredients easily available to people. One can be highly motivated and very well able to perform the target behavior, but still not perform the behavior. Triggers might be needed to remind people to perform the behavior. Three types of triggers are distinguished. The spark is to be used if ability is high and motivation is low, in order to inspire and motivate people. The facilitator is to be used if motivation is high and ability is low, for example by explaining to people the behavior is not difficult to perform. The signal is to be used when both the motivation and ability are high, but people need a reminder not to forget the target behavior.

For this project the ultimate target behavior can be defined as adopting an eating behavior that enables maintaining an appropriate weight level (see figure 6.2). This target behavior can be split up in three sub-targets, in line with the three design directions formulated previously: (1) dishing up a proper portion size, (2) slowing down eating rate, and (3) waiting a certain amount of time before taking a second serving. These three sub-target behaviors stimulate the user to eat more mindfully. As a result of mindful eating, users should feel more satisfied with and satiated by their food intake, which should prevent them from taking a second serving, and thus lead to weight management. If users decide to take a second serving after all, they will be

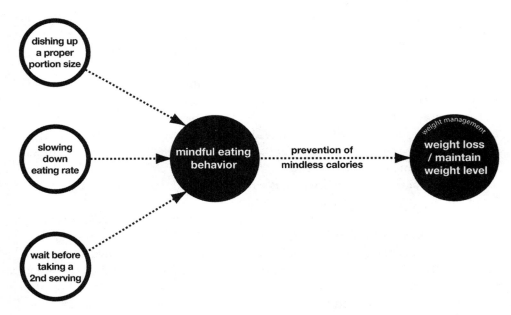

Figure 6.2
Three sub-target behaviors that should lead to more mindful eating and proper weight management.

Table 6.1

Analysis of three design directions by means of the behavioral model for persuasive design

	Portion size	Slow down eating	Second serving timer
Target behavior	Dish up the correct amount of the different food groups	Take time to consciously enjoy every bite	Wait at least 20 minutes after the start of your first serving before deciding whether to take a second serving.
Ability	Brain cycles (-/-) Social deviance (-/-)	Non-routine (-/-)	Social deviance (-/-)
Motivation	Want to lose weight Want to adopt a healthy lifestyle	Want to lose weight Want pleasure of improved taste experience	Pleasure of not overeating
Trigger	Facilitator	Signal (subtle and personal)	Facilitator spark

making a well-considered choice, not a choice sparked by ambivalent or subconscious feelings.

As part of the analysis by means of Foggs's behavior model, we investigated why people find it difficult to perform the target behaviors in their current situation and how a product could potentially help them. Four participants were asked to perform one of the three target behaviors for a week and report which positive and negative experiences they encountered. Table 6.1 summarizes the results of the analysis. There are different reasons why our participants found the target behaviors difficult to perform. People do not measure their food intake because it takes too much (cognitive) effort and it is not socially accepted to rigidly monitor your food intake: the ability is low. For slowing down the eating rate however, the ability and motivation is already quite high. People only need an appropriate trigger to remind them it is better to eat at a slower pace.

Idea Generation

For each of the three directions, we created two product ideas and usage scenarios that take into account the findings of our analysis combined with the behavioral model for persuasive design described in the section above. We describe these six ideas and evaluate their potential for end users in this section.

To support people in dishing up an appropriate portion size, we created the "serving spoon" and "chopping board" (see figure 6.3). Both concepts provide users with feedback on a healthy portion size and a balanced composition of ingredient groups (vegetables, proteins, grains, etc.) of their main meal. The serving spoon can weigh the food while you dish it up and keep track of the additional weight with every spoon.

Figure 6.3
The "serving spoon" (left) and "chopping board" (right) to support people in dishing up an appropriate portion size.

The concept consists of different types of spoons and spatulas, organized in a storage base. You can choose to weigh the uncooked food, before putting it in the pan, or to weigh the food when you put the meal on your plate. There are different spoons, depending on the size and type of the food. All spoons have a slider to select the food group, a tare button, and a rotating end to select whether the food is raw or cooked. When a spoon connects to the storage base, it transfers the data and saves it. The "chopping board" is a cutting board with integrated scale. Users can weigh the food while they are chopping it or when dishing up, by putting a plate on top of the chopping board, and the board gives feedback about the portion size.

In order to remind people to eat more slowly, we created "vibrating cutlery" and the "warming plate" (see figure 6.4). The vibrating cutlery detects when users move the cutlery too quickly, and it vibrates for a short period to remind users to eat at a slower pace. The warming plate keeps the food warm so people can eat at a slower tempo without the food getting cold. But when users eat too fast, the warming function will turn off. A light indicates whether the heating function is on, so users can easily assess whether they are eating too fast.

Finally, to motivate people to wait before taking second servings we created "timing cutlery" and the "timer box" (see figure 6.5). The timing cutlery consists of a timer made from silicone that users can slide onto the end of regular cutlery. It starts running when they take their first bite and gives a signal when twenty minutes have passed. The "timer box" can be used to store the leftovers directly after dishing up the first serving. When the timer box is closed, it starts counting down from twenty minutes

Figure 6.4
The "vibrating cutlery" (left) and "warming plate" (right) to help people eat more slowly.

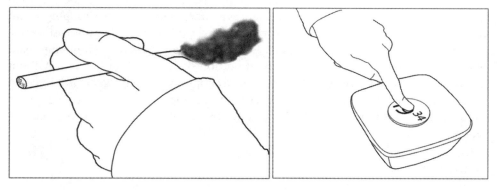

Figure 6.5
The "timing cutlery" (left) and "timer box" (right) to help people to wait before taking a second serving.

and keeps the food warm. The countdown is visualized by a timer. After twenty minutes, users can decide to eat the leftovers or to remove the thermo sleeve and put the box in the fridge for another day.

We evaluated the usage scenarios of the six ideas with twelve participants who had an interest in weight management. Based on the evaluation, it seems that there are two target groups for the six ideas. One target group consists of "weight managers" who realize they have to exert effort if they want to change their eating behavior. They want to invest money and time in order to manage their weight, but they also expect the product to be effective in assisting them to achieve their goals. This group was mainly interested in the chopping board idea. The second target group consists of "uncontrolled eaters" who are satisfied with their food intake but are interested in the means to control their eating rate. Losing weight will only be a bonus; it is not their intention. They were interested in a combination of the warming plate, vibrating cutlery, and timing cutlery.

Based on the user feedback, the chopping board concept for "weight managers" seemed to have the most potential. The chopping board aims to facilitate this food intake monitoring process with little cognitive effort from users. A chopping board was considered very appropriate for this task because it is used at the moment people are already concerned with how much food should be prepared. A chopping board offers the opportunity to give advice about portion size at the right moment without it being perceived as unsolicited. Furthermore, a chopping board is a generally accepted kitchen tool that is already part of the current cooking process for many people; it would not require the introduction of yet another appliance or tool in the cooking process and on the kitchen counter.

Conceptual Design

Although the participants positively evaluated the basic idea of a chopping board to facilitate the food intake monitoring process, we realized that design decisions critical to user acceptance, such as the material used and the way in which the data is presented, also needed testing. As it was not clear what the users' preferences were with respect to such details and how it would affect their usage of the product, we designed three different chopping board concepts. The concepts varied in several aspects, including size, interaction style, data display, materials, and aesthetics. We made mock-up models of every concept and evaluated them with participants in a kitchen environment.

The first concept consists of a thick wooden chopping board, which has a permanent place on the countertop (see figure 6.6). The product has a touch-screen interface at the top of the board. User data is stored on a user token, which is positioned at the side of the board by means of a magnet. The second concept is a more modern looking,

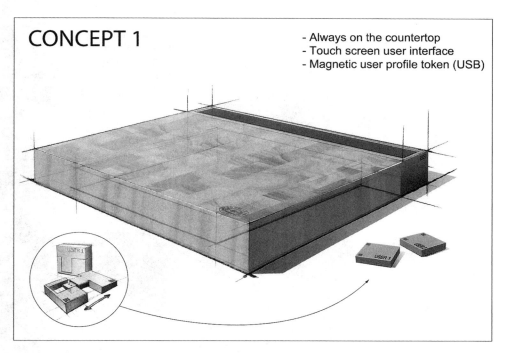

CONCEPT 1

- Always on the countertop
- Touch screen user interface
- Magnetic user profile token (USB)

Figure 6.6
Chopping board concept 1: pretty on the countertop.

thinner product with replaceable chopping boards of epicurean material (see figure 6.7). There are no physical buttons; the controls on the product work by means of capacitive sensing. Each user has a physical cube, which is used to control the device and show a user's individual results. The third concept is a very thin, basic, and lightweight plastic chopping board that is easy to store in a cupboard, like a regular chopping board (see figure 6.8). It has physical buttons for user input with tactile feedback. A small user token is positioned at the top side of the chopping board. The user can position different cutting sheets on top of the product, to prevent cross contamination between meat and vegetables, for example.

Five female and three male participants evaluated the three mock-up models. All participants had an interest in losing weight, weight management in general, or in healthy eating behavior. Almost all participants had a one- or two-person household and differed in age from early twenties to mid-fifties. Six out of eight participants were employees of the industrial research organization in which this project was conducted, but none were involved in a project related to the topic of mindful eating and behavior change.

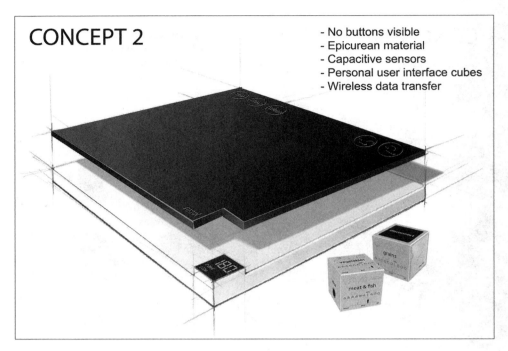

Figure 6.7
Chopping board concept 2: divide and conquer.

The evaluation of the three chopping-board concepts made clear which design aspects were considered important for potential end users and which solutions were preferred. Seven participants liked concept 3 best because of its simplicity, the fast and unambiguous interaction, its compact dimensions, and because it is easy to clean and store. Although concept 1 with the wooden chopping board was considered the most beautiful, participants preferred the plastic chopping board of concept 3 for functional reasons, including hygiene, easy storage, and ease of use. Concept 2 was disliked by the participants mainly because the interaction with the cubes was considered too complicated.

Detailed Design and Functional Prototype

Based on the results of the concept evaluation, we developed a more detailed design. The "InBalance" chopping board includes a scale in a high-quality plastic casing, on which thin chopping sheets are placed. The sheets can easily be cleaned in a dishwasher and be replaced if they wear out. Since proper portion size depends on user

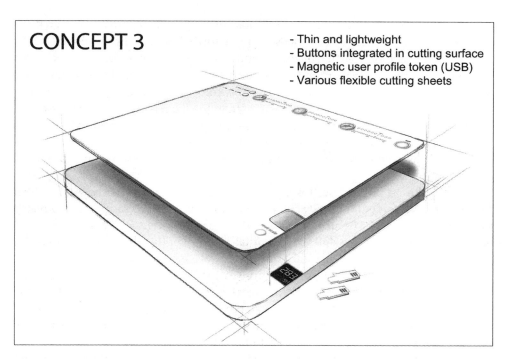

CONCEPT 3

- Thin and lightweight
- Buttons integrated in cutting surface
- Magnetic user profile token (USB)
- Various flexible cutting sheets

Figure 6.8
Chopping board concept 3: slim as an ordinary chopping board.

characteristics such as gender, age, weight and height, and physical activity, the scale needs to know who the current user is. Therefore, users can connect a token that stores their profile on the scale. After identifying themselves with this token, users indicate whether the ingredient that is placed on the scale is cooked or uncooked. Furthermore, they specify to which food group it belongs (vegetables, meat/fish/meat replacements, or grains) by pressing buttons on the board. Since the technology to automatically detect ingredients has not been fully worked out—in particular for fresh ingredients that have no barcode or tag—we decided to adopt this user involvement in identification. In addition, users have to indicate whether the ingredient is a food product that one can healthfully eat always ("go," as with vegetables), sometimes ("slow," such as chicken with skin), or only very rarely ("whoa," such as pork chops) (CATCH 2002). This categorization is based on nutritionists' recommendations. Although the method for ingredient categorization seems shallow, we deliberately chose it in lieu of a more detailed ingredient specification for a number of reasons. First, we expected that users would find this categorization easier to work with. Second, it would make the interaction with the chopping board easier and faster. Third, based on discussions with nutritionists, it appeared to be a very workable and functional approach to ingredient classification. As a result of the user's input, a row of light-emitting diodes (LEDs) indicates the share of the ingredient currently on the board compared to the target portion size of that particular food group for the current user. The LEDs representing this "portion size score" are white, because this color is neutral and does not refer to "good" or "bad." Only the LED signaling that the target portion size is reached becomes green (to indicate the ideal). The user can press the "save" button to save the portion size score onto the user token. The system currently allows up to two separate tokens to be used with the board, so meals for for up to two persons can be prepared at the same time.

The data stored on the user token can be uploaded to a personal website, which presents an overview of the users' achievements and coaching, keeping users motivated to reach their goals. The homepage of the website provides users with a quick overview of yesterday's achievements and the food intake history of the current week. Users can select the button labeled "food intake history" to get a more detailed overview of what they've eaten on a daily, weekly, monthly, or yearly basis. The website can also support people by providing healthy recipe recommendations (see also chapter 3 by Jettie Hoonhout et al.). The recommendations can range from a suggestion for a new healthy recipe when the user asks for inspiration to a complete week's menu with healthy recipes tailored to people's needs and diet plans. The variation balance (see figure 6.9) gives users insight into the variation in their food intake, based on the "go," "slow," "whoa" categorization mentioned above. Furthermore, lists of example ingredients for each of the categories can be viewed. Another section of the personal coaching website provides tips for mindful eating. Finally, while the chopping board assists in monitoring the food intake on a global level (i.e., amount of meat, grains, vegetables), the website provides the option for more accurate monitoring. Users can select the exact ingredients and the preparation methods of the ingredients they have weighed on the chopping board. They also have the option to enter other food intake data, for example, snacks, lunches, and breakfast. The coaching application can give more accurate and detailed advice based on the more complete input.

User Evaluation of Final Concept

The final concept of the InBalance chopping board was evaluated in the kitchen of ExperienceLab. ExperienceLab is a research facility to conduct end-user research in a realistic but controlled environment during early stages of product innovation. The ExperienceLab has a fully functioning kitchen where our participants prepared a meal and used the functional prototype of the chopping board. This facility was considered particularly useful for our research, as we are interested in how people experience the interaction with the chopping board while they are cooking. The main focus of the

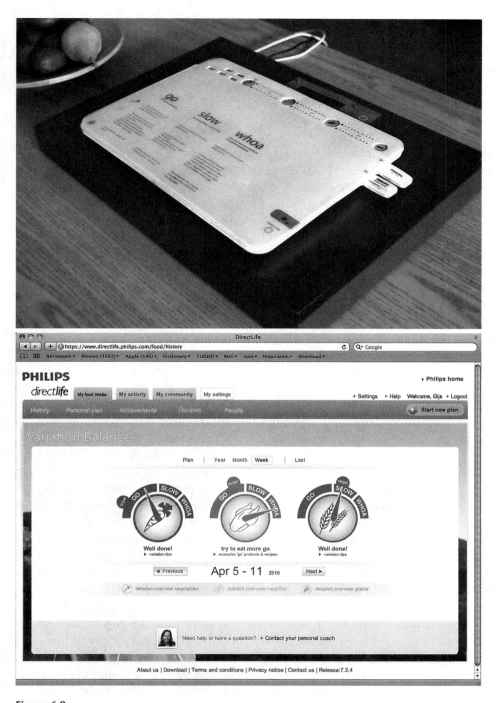

Figure 6.9
Functional prototype of the InBalance chopping board (top) and a screenshot of the variation balance on the personal website (bottom).

test was on detecting short-term usability and user interaction issues and participants' attitude (based on first impressions) toward the function and perceived effectiveness of the product.

In total, we recruited ten participants (four women, six men) to individually prepare a meal using our product concept. They were selected based on their interest in losing weight or weight management and had to cook regularly. They lived in a one- or two-person household and were between twenty-five and sixty years old. Nine participants were Dutch and one participant had lived in the Netherlands for over ten years. Prior to the evaluation, the participants received information about the "go," "slow," "whoa" classifications and received a copy of the recipe they had to prepare. The evaluation itself consisted of three parts. In the introduction, the experimenter explained the background of the project and demonstrated how to use the prototype. Next, the participants were instructed to prepare a recipe (rice with stir-fried vegetables). They had to use the prototype to measure the vegetables and meat before cooking them and the grains after cooking them. During this part of the test we observed and recorded the participants from a separate room by means of video cameras. Afterward, we asked participants for their opinion about the prototype and its use. In a semistructured interview, participants answered questions about a range of topics including their first impressions, the interface, the physical embodiment, and the usability of the chopping board.

Overall, the participants had a very positive attitude toward the product concept. They all mentioned that using it did not take much (cognitive) effort, and they did not perceive its use as obtrusive during the cooking process. The following remarks made by our participants illustrate the positive feedback above: "It was very clear actually . . ." (participant 9); "I like it! Very easy. It did not take any extra time. Ping, Ping, done!" (participant 8). Although participants agreed it might be difficult at the start to learn which ingredients are "go," "slow," or "whoa," they did not perceive this as a problem. To the contrary, many liked this aspect of the product, because it confronts them with their food choice and enables them to make better choices the next time they are buying food. Example ingredients for each category are printed on the board (the blocks of text printed on the board, see figure 6.9) to help users get started. The participants indicated that they thought that the product would keep them motivated to learn more about healthy and less-healthy ingredients. As one participant phrased it: "If you are interested in mindful eating you don't mind having to learn which products are 'go', 'slow,' or 'whoa'" (participant 8).

Interestingly, the participants were highly focused on getting the amount of food on the board precisely right such that the LEDs would show a 100 percent portion size score, regardless of the food category. However, for most "go" ingredients it is actually not an issue to eat more than the indicated 100 percent. Only for more calorie-intense ingredients, such as meat, is it not advised to exceed 100 percent. Another interesting finding was that participants had different motivations for either wanting to weigh raw ingredients or cooked ingredients. The main reason for measuring raw ingredients was to prevent cooking too much and wasting food. On the other hand, participants suggested that weighing cooked foods would be helpful when preparing food for several days.

The participants experienced few usability issues during the test, although some initially forgot to press the save button after weighing some ingredients. In these occasions, the blinking LED light helped them to correct that omission. Only one participant did not save the results at all. Overall, the participants thought that learning how to interact with the device was easy, certainly after one trial.

Conclusion and Discussion

Following a user-centered design approach, we investigated how to make people more mindful about their eating behavior during dinner time in order to persuade them to adopt a healthier lifestyle. We explored three design directions that can influence the consumption of excessive, mindless calories during dinner. The first direction focused on supporting people to eat appropriate portion sizes. The second and third design directions focused on creating awareness of satiety level, either by slowing down people's eating rate or by making people wait longer before deciding whether or not to take a second serving. For each of the three directions, two product ideas and usage scenarios were created, taking into account the guidelines based on Fogg's behavioral model for persuasive design (2009). These six ideas were evaluated with potential end-users, leading to the decision to further develop the chopping board concept. After a user evaluation of three different conceptual designs, we developed a final design and working prototype of the chopping board.

The resulting InBalance chopping board is a cutting board with an integrated scale that gives feedback about portion size. This final design was evaluated in a realistic test setting in which people prepared a meal using the prototype. Overall, the participants had a very positive attitude toward the product concept and no major usability problems were identified.

The results suggest that the InBalance chopping board could be effective in helping users eat more mindfully in order to manage their weight by increasing people's awareness and knowledge of healthy ingredients and portion sizes. The participants perceived the concept as easy to use and easy to integrate into their current cooking behavior, allowing them to better control their portion sizes during dinner. However, our final evaluation mainly focused on users' initial reactions to and interactions with the product concept. The next stage of evaluation should address the long-term effects and effectiveness of our concept in changing people's lifestyles. Furthermore, it is important to note that this study was very much tailored to the Western European (i.e., Dutch) food culture. It is likely that opinions of people in other food cultures about mindless eating and the InBalance chopping board will be very different.

We believe that a process in which potential end users are involved systematically and during various stages of the process to give feedback about the design leads to better end results. Although it is impossible to capture the valuable input of potential end users completely, given the amount of data and its richness, we tried to illustrate in this chapter how participants' feedback can guide the design process into a certain direction. In our opinion, the positive feedback of end-users on the InBalance chopping board strongly indicates the success of a user-centered way of working.

We believe technology can play an important role in supporting people to change their lifestyle in a positive way and prevent lifestyle-related health problems in today's and tomorrow's societies. The InBalance chopping board is just one example of a product that could help people in living a healthier lifestyle. We can think of many more: for example, products that help people to increase physically activity, to reduce snacking, or to sleep better. The chopping board is only a small step toward a broader range of (probably connected) solutions that can support people in a healthier and more balanced lifestyle.

Acknowledgments

Our research was partially funded under the SmartProducts project (EC-231204).

We also thank Jasper van Kuijk and Daan van Eijk of the Faculty of Industrial Design of the Delft University of Technology for their contribution to this project.

SPICY SHRIMP AND VEGETABLE STIR-FRY, USING THE INBALANCE CHOPPING BOARD

Makes 4 servings

Nutritional value per serving (not including the rice or noodles): 301 calories, 6.6 g. fat (1 g. saturated), 24 g. carbs, 3.3 g. fiber, 32 g. protein; this would be a "go" dish.

Stir-frying vegetables is a quick way to prepare a tasty and healthy dinner. When done correctly, stir-frying uses minimal fat/oil and leaves the vegetables bright, colorful, and tender-crisp. It also uses little to no water, so vegetables will lose fewer water-soluble nutrients such as vitamin C. Stir-fried vegetables can be served with rice or noodles.

Stir-frying foods is usually done in a wok. The wok's bowl shape requires little oil compared to other pan shapes. The vegetables in a stir-fry should be cut so they cook in roughly the same time, with dense vegetables such as carrots cut thinner than bell peppers or broccoli.

Ingredients

¼ cup low-sodium soy sauce

¼ cup sake

2 tbsp. sugar

1 tbsp. dark (toasted) sesame oil

1 tbsp. chopped garlic

1 tbsp. finely chopped or grated ginger

1 cup large-diced red bell pepper

1 cup large-diced green bell pepper

1 cup large-diced onion

1 cup cubed cabbage

1 cup sliced carrot

½ tsp. red pepper flakes

24 large shrimp, shelled and deveined

Preparation

1. Combine first 6 ingredients in a bowl. Heat a large nonstick wok over medium-high heat. Add soy sauce mixture; cook for about 1 minute, until slightly thickened.
2. Add vegetables and red pepper flakes. Cook, stirring constantly, until vegetables are soft. This will take just a few minutes.
3. Add the shrimp and 1/4 cup water. Cook, stirring often, until shrimp are just cooked through (approx. 2 to 3 minutes). Divide among 4 bowls and serve immediately with steamed rice or noodles.

References

Baumeister, Roy F., Ellen Bratslavsky, Mark Muraven, and Dianne M. Tice. 1998. Ego depletion: Is the active self a limited resource? *Journal of Personality and Social Psychology* 74: 1252–1265.

CATCH. 2002. What is GO, SLOW and WHOA. http://catchusa.org/programs.html.

Fogg, B. J. 2009. A behavior model for persuasive design. In *Persuasive '09: Proceedings of the 4th International Conference on Persuasive Technology*, New York: ACM.

Jansen, Anita, Chantal Nederkoorn, Anne Roefs, Carolien Martijn, Remco Havermans, and Sandra Mulkens. 2009. Waarom obesitas in de GGZ behandeld moet worden (Why obesity should be treated in mental health care facilities). *GZ-Psychologie* (December 2000):38–44.

Kramer, F.Matthew, Robert W. Jeffrey, Jean L. Forster, and Mary Kaye Snell. 1989. Long-term follow-up of behavioural treatment for obesity: Patterns of weight regain among men and women. *International Journal of Obesity* 13:123–136.

Van Buren, Dorothy, and Meghan M. Sinton. 2009. Psychological aspects of weight loss and weight maintenance. *Journal of the American Dietetic Association* 109 (12):1994–1996.

Wansink, Brian. 2004. Environmental factors that increase the food intake and consumption volume of unknowing consumers. *Annual Review of Nutrition* 24:455–479.

Wansink, Brian. 2006. *Mindless eating: Why we eat more than we think*. London: Hay House.

Woods, Stephen C. 2004. Gastrointestinal satiety signals I: An overview of gastrointestinal signals that influence food intake. *AJP—Gastrointestinal and Liver Physiology* 286:G7–G13.

7 Encouraging Fresh Food Choices with Mobile and Social Technologies: Learning from the FlavourCrusader Project

Grant Young and Penny Hagen

A lone strawberry farmer at a bric-a-brac market planted the seed for FlavourCrusader. His strawberries were amazing—oozing with sweetness and a depth of flavour I had never experienced. The strawberries were soft; their colour dyed my fingers. I told my friends. I emailed others. I blogged it. Go to Rozelle markets for the strawberry man!

I had gotten out of the habit of eating fruit after living in London. It was tasteless. It was pointless. The delicious strawberries were the catalyst for me to begin eating fruit again.

Deliciousness can be a catalyst for change.
FlavourCrusader founder Sharon Lee (Lee and Williams 2011)

Millions of people now have access to a variety of fresh food from all over the world due to the advance of the global industrial-commercial food system. Many of us no longer experience the restrictions imposed by local environmental factors or seasonality (Gaballa and Abraham 2008). Indicative of this is the significant growth in major grocery retailers' fresh food sales (McKinna et al. 2007), which currently account for between 20 and 25 percent of grocery sales in Australia (PricewaterhouseCoopers 2007).

These developments have not come without cost. Concerns about this globalized model's impact on environmental sustainability, animal welfare, health and safety, and socioeconomic issues, as well as a desire for improved quality and taste, are all driving interest in alternative food systems (Coster and Kennon 2005; Hendy 2010; Hogan and Thorpe 2009; Lockie 2008; Meat and Livestock Australia Limited 2008; McKinna et al. 2007; Vermeir and Verbeke 2006; Winter 2003). Such concerns have also contributed to the promotion of local food production models (Department of Agriculture, Fisheries and Forestry 2006; Manzini 2007; Winter 2003). Over 150 registered farmers' markets now operate across Australia (Australian Farmers' Market Association 2010) in service of this growing market. Consumer demand for organic produce has also seen strong growth, with 70 percent of organic food consumers purchasing at least once a week and approximately 60 percent of households buying organic produce on occasion (Hendy 2010).

FlavourCrusader is a volunteer-led initiative with participants (including the co-author of this chapter, Grant Young) emanating from the social innovation movement in Sydney, Australia. The initiative is inspired by the premise that increasing the consumption of local fresh produce can contribute to:

• Improved health outcomes.
• Reduced transport impacts of food production.
• Increased food security.
• Socioeconomic benefits, specifically supporting local producers in the context of the high concentration of the Australian grocery retail market (PricewaterhouseCoopers 2007) and competition from food imports (McKinna et al. 2007).

Mobile and social technologies have the potential to allow people to access information about food at the point of purchase and decision making, to connect with local food producers, and to share their food experiences and knowledge with their peers and online acquaintances (e.g., "followers" or "friends"). The FlavourCrusader initiative investigates how such practices might be utilized to effect behavior change around the consumption of fresh produce and the promotion of locally produced fruits and vegetables in particular.

Behavior change models such as the "five doors" framework (Robinson 2011) and the Fogg behavior model (Fogg 2009) emphasize the need to understand people's *motivation*, enhance their *ability* to implement changes, and to understand what might act as a *trigger* for such new behaviors (Fogg 2009). Wanting to better understand these aspects in the context of promoting consumption of locally produced food, members of the FlavourCrusader team undertook a series of research activities to increase their understanding of what motivates people to choose local produce and the potential role that mobile and social technologies might play in supporting such choices. This included interviews with people already engaged in the consumption of locally grown produce, along with the prototyping and testing of an Australian seasonal produce guide for mobile devices.

This chapter outlines the findings from the initial interviews and how these findings led to the development of a seasonal produce guide. We then describe the process of developing and testing the guide and what the results of this testing suggested about how people make fresh-food choices. We close the chapter by reflecting on what both the interviews and early testing of this prototype have shown us about the potential for mobile and social tools to support better food choices. We also suggest potential steps for future research and product development.

It is important to note that the volunteer nature of the FlavourCrusader project means our approach to research, development, and testing is necessarily opportunistic:

we take advantage of the resources and occasions for participation and feedback we have at hand. While this impacts the scale of the initiative and our research, the interest and momentum for the project has come from a grassroots and community level; the FlavourCrusader project in itself is representative of the emerging interest in food and sustainability that this book reflects.

Buying Local

While there are numerous reasons *why* an increase in local fresh food production and consumption is beneficial, exactly *how* to make this happen is a more complex proposition. Market research undertaken for the Department of Primary Industries Tasmania indicates that the buying of vegetables is, for the majority, a "low-involvement activity" (McKinna et al. 2007). Cost, convenience, and quality are generally recognized as significant influencers of purchasing behavior, but for the most part there is actually very limited consideration given to purchasing decisions by consumers.

In an effort to understand the motivations of those who do buy local we conducted semistructured interviews with five "social media–savvy foodies"[1] and two people engaged in alternative food production. Interviews were between sixty and ninety minutes in length. The interview questions explored three aspects in particular: the role food played in their lives; motivations for purchasing local produce; and the current and potential role of mobile and social tools in relation to food buying and cooking. The first of these aspects was documented and translated into a set of personas used to support communication about the project (available at FlavourCrusader Personas, Young 2010a). The second and third aspects more directly informed the development of the seasonal produce guide prototype and are described in more detail below.

We found there to be a range of motivations for buying local produce, including:

• *Connecting with food producers*: knowing where and how food was grown was important to research participants, contributing to their sense of community and concomitant trust.

• *Supporting the local economy*: a desire for more of their money to reach producers and stay within producer communities was also evident. This was tied to a sense of social justice and equity for some participants, specifically that producers receive "fair" prices for their produce.

• *Improved taste and quality*: there was a perception that local fresh produce purchased through non-mainstream retail channels was fresher, of higher quality, and more flavorsome.

• *Health benefits*: participants assumed that local fresh food sourced through non-mainstream merchants had higher nutritional content and reduced chemical inputs.[2]

• *Sustainability*: local produce was considered more sustainable due to lower "food miles," and therefore lower total emissions,[3] than imported produce. Participants reasoned that major grocery retailers' distribution models (e.g., produce delivered via regional distribution centers) also resulted in higher food miles than produce purchased through non-mainstream retail channels. The perceived reduction in chemical inputs was also considered a benefit from a sustainability perspective (in addition to the health considerations mentioned above).

• *Distrust of mainstream retailers and certification schemes*: there was a lack of trust that major grocery retailers, such as Coles and Woolworths, were meeting expectations in relation to the factors outlined above. Participants also expressed distrust in certification schemes, especially organic certification.[4]

The appeal of local fresh produce, then, was driven by diverse social, ethical and environmental motivations, as well as an expectation of fresher (better quality and tastier) food. While our research represents the views of a small and quite specific sample of interviewees, these findings correlate with existing literature on motivations for buying local produce (see, for example, Cooler Solutions Inc. 2011; Meat and Livestock Australia Limited 2008; Vermeir and Verbeke 2006; Winter 2003). For our interview participants, already motivated by ethical, social, or environmental factors, buying local was a heuristic or shorthand for dealing with a number of potentially separate and quite complex domains relating to food production and consumption.

At the same time, there were several barriers identified by participants in relation to buying local produce, especially from farmers' markets. Among other concerns, knowing what was in season was a key information gap for participants, especially when purchasing produce at major grocery retailers. In environments such as farmers' markets there was an implicit expectation that what is on sale is in season, while in mainstream supermarkets it was seen as more difficult to determine whether food on display is in season or not.

In exploring the role that mobile and social tools played in their food-related activities, participants noted using recipe sites and applications, "check-in" services such as Foursquare and Facebook, and rating/review tools like UrbanSpoon (UrbanSpoon 2011) for evaluating local prepared food outlets (i.e., eateries and restaurants). Some also indicated that they engaged with popular social networks such as Twitter for sharing food-related photos and status updates. Examples of such activities included sharing the excitement of purchasing special fresh produce, exchanging recipes and cooking tips, or discussing industry-related news. The potential to connect to producers and other "foodies" also featured strongly for some participants. While farmers' markets provided an opportunity to create such connections, these connections were also augmented or facilitated by online social networks. The interviews also identified specific gaps where new tools might support the purchase of fresh local produce, including:

• Support for decision making about what to eat and cook in situ (e.g., at point of purchase).
• Information that would increase people's ability to determine what was in season, particularly when shopping at mainstream outlets.
• Localized content, as existing tools were predominantly United States–centric.

While a number of potential design directions presented themselves as a result of this initial research, the FlavourCrusader team focused on developing a mobile application that could provide information about seasonality of produce. At the time the availability of such information was limited, especially in a portable format. While printed seasonal food diaries and recipe books were available, mobile applications such as the Seasons iPhone application (What Is It Productions Ltd 2010) did not include Australia-specific data. An Australian mobile seasonal produce guide addressed the need for localized food information and had the potential to support decision making during meal planning and/or at the point of purchase.

Importantly, a focus on *seasonal* (rather than *local*) produce afforded the ability to encourage the purchase of locally produced food in contexts beyond the local farmers' market. Emphasizing seasonality enabled us to address more widely shared purchasing motivations. Seasonal produce is often cheaper, fresher, and can reduce waste through less spoilage (an identified barrier to the purchase of fresh produce [Brand Story and Horticulture Australia 2010]). Thus, this focus provided a way to promote local fresh produce to a larger audience than just those motivated by ethical, social, or environmental concerns.

Prototyping a Mobile Australian Seasonal Produce Guide

An initial prototype—a web application[5] suitable for iPhone and Android-based smartphone devices—was developed to test the effectiveness of a mobile seasonal produce guide in achieving the initiative's goals.

The prototype was first tested by the FlavourCrusader team and then across two workshop sessions with approximately forty participants. The initial feature set consisted of a listing of fresh produce currently in season, with a brief description of each item, a photo, and additional information on storage and preparation. Testing by the FlavourCrusader team quickly identified that listing *only* seasonal produce was not that useful in the context of trying to prepare a meal or evaluate a recipe. In response, an "All Produce" tab was introduced that provided a full list of all produce in the application's database, with seasonal status visually indicated. Some team members were not familiar with how to use less-common produce items listed in the application. This prompted the inclusion of a "Recipes" tab, which housed a rudimentary list of recipes incorporating in-season produce. By providing further information about how an item

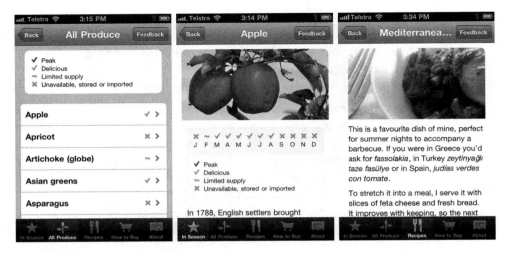

Figure 7.1
Screenshots illustrating (from left to right) the "All Produce," the "Produce Item" detail, and "Recipe" detail screens in the third (final) prototype.

could be incorporated into meals, the team hoped to encourage the purchase and use of less-understood produce items.

The second iteration of the prototype, which incorporated the "Recipes" and "All Produce" tabs, was then tested and further evolved across the two workshop sessions. The first workshop took place on February 26, 2011, at a BarCamp (Wikipedia 2012) organized by Social Innovation Sydney (Social Innovation Sydney 2011). The second workshop with students from the RMIT University's Master of Sustainable Practice program was held the following week. Figure 7.1 shows three screens from the application as implemented for the second workshop.

Voluntary attendance at the host events was the only criteria for participation in these workshops. While no demographic or other data (e.g., participant attitudes or attributes) were collected during these sessions, the context of the workshops implies that participants had a greater interest in social innovation and sustainability than would be present in the general population.

Setting up the Mobile-Use Tests
The personal nature and scale of mobile devices and the diverse range of places in which they are used makes testing mobile technologies highly problematic. Simulations of real-use contexts in controlled environments are one way in which researchers have worked to provide access to information about potential mobile usage (Hagen et al. 2005; Kjeldskov and Skov 2007). To support testing of the prototype with workshop participants, we created three different stations. Each station simulated a

different mobile-use context corresponding to a potential situation where a seasonal produce guide might be of use in the preparation and cooking of a meal. The terms "fresh" and "delicious" were used to frame the scenarios in relation to the end-user benefit (a tasty meal), rather than the function of the tool (seasonal produce). The scenarios were:

1. You are on the bus home from work and you want to cook something delicious for dinner.
2. You are planning a dinner party with friends on the weekend and you are working out what you will cook and need to buy.[6]
3. You are at a store and you want to get fresh, delicious produce.

The first station consisted of three rows of seats to mimic traveling in a bus (Scenario 1: Bus); another emulated a kitchen table with recipe books and a shopping list (Scenario 2: Dinner party); the third represented a supermarket display with a selection of fresh fruit and vegetables, both in- and out-of-season (Scenario 3: Store—see figure 7.2). A volunteer facilitator was available at each station to answer questions related to the testing activities and to guide discussion. Participants were asked to self-organize in approximately equal-size groups across the stations. They were instructed to

Figure 7.2
The "Store" station at Social Innovation Sydney BarCamp, February 2011 (photo courtesy J. J. Halans).

consider how they would approach the scenario using the seasonal produce guide for support. Participants utilized their own smartphones for the exercise, with a majority of participants using iPhones. Two participants utilized Android-based smartphones and the few participants who lacked a suitable device shared with participants who had one.

In keeping with the format of the Social Innovation Sydney event, the initial test session length was thirty minutes. This was extended to forty-five minutes for the RMIT University workshop. Feedback was collected through self-reporting sheets with a short number of questions related to each scenario. An "ideas board" flipchart, with a supply of Post-it notes and pens, was also provided for capturing less structured feedback such as feature suggestions. A short discussion occurred at the end of each session, with the lead facilitator asking unstructured questions of the group and collecting further feedback. This broader discussion also served to uncover opportunities for the application to provide proactive prompts, or triggers, that might promote fresh produce purchasing and consumption within the guide (something the FlavourCrusader team saw as important for the future development of the application).

The prototype was iterated for the second workshop to include a "How to Buy" tab displaying a text-based list of farmers' markets organized by state.[7] The introduction of this feature was prompted by participant requests for such a feature in the first workshop. Minor changes to the layout and presentation of information were also implemented in this third (and final) prototype iteration.

What We Learned

In testing the prototype we aimed to evaluate the role a mobile seasonal food guide might play in different scenarios of use. We also sought to identify other potential ways in which social or mobile tools might support or encourage participants to buy seasonal produce. The testing highlighted the importance of understanding and designing for users' existing approaches to meal planning. In response to the three scenarios, three distinct pathways emerged that provided insight into how people approach planning a meal in different situations, including how they make decisions about what fresh produce to buy and where to buy it. These pathways (described below) made it clear that providing information about what is in season alone was insufficient to encourage people to purchase local or in-season ingredients. However, they highlighted opportunities to promote local produce through seasonal recipes and helping people determine where to buy their local fresh produce as part of meal planning. The testing also provided further insight into the potential triggers that could encourage different purchasing behaviors. Lastly, the importance of performance of mobile tools supporting in situ decision making also became apparent. Each of these aspects is explored in further detail below.

Pathways to Planning a Meal

Figure 7.3 documents our interpretation of the pathways to meal planning and purchasing that workshop participants described when considering the test scenarios. Importantly, the act of identifying or selecting seasonal ingredients is only one step of many involved in planning a meal. Participants also considered what they would like to cook, what would accompany any primary dish or ingredient, and where produce or ingredients could be purchased. Scenarios 1 (Bus) and 2 (Dinner party) were likely to begin with either a theme (e.g., Italian) or cooking style (e.g., stir-fry). The choice of recipe may influence this choice, or follow closely. Planning meals usually started with the primary ingredient, typically the main protein source. For non-vegetarians this is often the meat around which the dish is arranged (a finding supported by Meat and Livestock Australia Limited 2008). Available fresh produce therefore is not a common start point for meal planning. Scenario 3 (Store) is the only pathway where purchase decision making starts with consideration of specific/available fresh produce ingredients. In this case the appraisal of available ingredients may include considerations of seasonality, but only among other attributes (such as perceived freshness of produce and price). Consideration of recipes and how to use the ingredient came later in the process in this scenario.

These findings suggest that information about seasonality represented at the level of the recipe, rather than ingredient, may be more effective in supporting people to

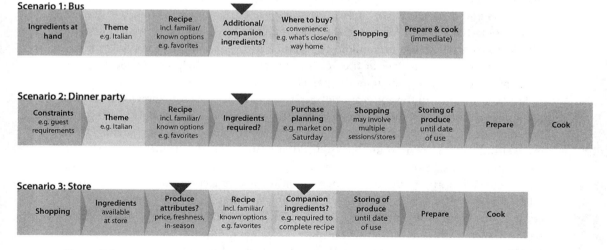

Figure 7.3

Three pathways to preparing and cooking fresh produce. Light grey items are optional. Arrows indicate where consideration of seasonal produce is likely to occur.

change their patterns of consumption. Such an approach, therefore, may be more effective in prompting people to change their patterns of consumption.

Seasonal Recipes

The scenarios demonstrated that participants might be more likely to purchase seasonal ingredients if the ingredients were promoted at a meal-planning level, for example as part of a theme or recipe. To achieve this, participants suggested the following improvements to the rudimentary recipe listing included in the prototype:

• Updating the ingredient detail page to include a list of recipes that use/feature the ingredient.
• Explicitly indicating the seasonality of recipes (this was only implied in the recipe list implemented for the prototype).
• Filtering recipes to display those that consisted entirely of in-season ingredients.

Emphasizing recipes could also support the introduction of new or unfamiliar produce items into the users' meal-planning process, potentially expanding seasonal buying habits. A number of interview participants had shared stories about discovering novel or unfamiliar items at farmers' market but then being challenged to work out how to cook or use the item. There are multiple ways to overcome this challenge: at a farmers' market there is an opportunity to talk to the producer directly; existing skills and knowledge may point to how an unfamiliar item can be utilized; social networks may be tapped for ideas or tips. Our interpretation of participant comments in both interviews and during the workshops is that a sense of self-efficacy (ability) is needed to overcome the sense of risk associated with purchasing unfamiliar produce. Recipes can demonstrate how the unfamiliar item can be incorporated into a complete meal. Thus the provision of recipes at the point of purchase could help people lacking this sense of self-efficacy to overcome the perceived risk of purchase.

Shifting the focus of the application to recipes is not without its challenges, however. A number of social and mobile applications that provide original recipes, or support the sharing of recipes, already exist. For example, Epicurious (Condé Nast Digital 2011) is a website and companion mobile application that leverages a strong community of recipe sharers. A significant proportion of the value of tools like Epicurious resides in an engaged community supporting them. Building a strong community of this nature takes significant time and effort and, as with any site that is driven by user-generated content, the quality of submissions (and subsequent utility) is not guaranteed. Other recipe applications, such as those developed for Jamie Oliver's brand (Jamie Oliver Enterprises 2011), leverage the strong draw of celebrity for promotion. While few, if any, of these is focused on seasonality, they represent a significant degree of competition that needs to be considered.

Potential exists to tap into and leverage existing social technology platforms and their communities rather than providing recipes functionality directly. For example, it may be more efficient to utilize existing services providing recipe content, via an Application Programming Interface (API) for example. Leveraging Google's "Recipes Search" feature (see Singel 2011), or engaging a third-party recipe site such as AllRecipes.com, are examples of this approach. Alternatively, the tool might provide the facility to link to and associate existing recipes to seasonal produce. Providing a Delicious.com–style bookmarking tool that facilitates linking of recipes to a seasonal produce list is one potential approach.

Buying Fresh Local Produce

Participant requests for, and positive feedback on, the "How to Buy" feature highlighted its potential to support people in incorporating fresh produce into meal planning. The scenarios also highlighted ways to extend this capability depending on the different contexts of use. The Bus scenario (Scenario 1)—where convenience of access was a strong driver in decision making—demonstrated the value of being able to locate relevant merchants in close proximity or "on the way home" using geo-targeting features. Participants expressed a very strong expectation that such information would be presented in the form of a map. Interest in extending this functionality from farmers' markets to include local fresh produce merchants and prepared food vendors, such as restaurants that incorporated seasonal produce into their menus, was also expressed.

The additional available time implicit in a scenario such as planning a dinner party (Scenario 2) suggests a greater opportunity for merchant selection outside of typical buying patterns. That is, it is easier to plan to visit a farmers' market or local fresh produce retailer to purchase produce for a meal when there is more time between meal planning and purchase. In this scenario there is added utility in displaying a broader range of potential outlets, not necessarily just those closely related to the user's current location. Testing suggested that the "How to Buy" feature represented an opportunity to more actively guide users towards alternative channels for purchasing local produce.

In considering the expansion of the "How to Buy" feature it becomes apparent that the data sourcing and maintenance requirements for building the database of stores selling local produce are significant. Taking a crowdsourcing[8] approach, such as that employed by restaurant review applications such as UrbanSpoon and Beanhunter (Beanhunter 2011), is one way this burden may be addressed. However, the potential benefits of crowdsourcing need to be balanced with the prospect of abuse (e.g., false claims by merchants or astroturfing[9]) and would require a number of checks and balances, ongoing monitoring, and community management.

Another aspect of deciding "Where to Buy" highlighted by testing was in connection to price. A number of test participants suggested that they would decide on what and where to buy based on a price comparison of *individual* produce items (e.g., the price of broccoli or zucchini). However, maintaining a database of price data faces arguably greater challenges to those mentioned above, and its value in practice would need to be further evaluated.[10]

Potential Triggers

In addition to aiding existing meal planning practices, discussions with participants revealed the potential to prompt new purchase behavior through contextually relevant promotions. For example, alerts about local retailers offering good quality/prices or stores advertising discounts and specials could be delivered to participants via the mobile application. Despite the commercial orientation of such prompts, these promotions could also play a role in achieving FlavourCrusader's objective of encouraging seasonal produce consumption and are worthy of further examination.

Such triggers could include geo-targeted, time-based discounts and other promotions (e.g., "on special, today only"), Groupon-style group buying offers (e.g., "if 10 people commit, we'll give you a discount"), or the actions of friends in rating/reviewing a local vendor (e.g., "Your friend just rated a nearby grocer"). Employment of the "appointment" game dynamic (Priebatsch 2010; Schonfeld 2010) also holds potential for creating effective triggers. Creating a regular (weekly/monthly) "event" (similar in intent and form to "Meatless Monday" [The Monday Campaigns Inc. 2011]) or promoting seasonal ingredient "challenges" (reflective of the "challenge ingredient" activities of popular cooking shows such as Iron Chef and MasterChef) are two examples of such an approach.

Importance of Performance

The scenario testing also revealed other important technical factors to consider when developing applications for use in mobile contexts. For example, the Store scenario (Scenario 3) illustrated that the guide needed to load near instantaneously to be effective. Our prototype web application did not load and respond quickly enough to be useful at the point of purchase in a real-world context. A native application, with a local data store, has the potential to be significantly more responsive than is possible with a web application[11], and therefore a better choice for this type of application.

In Closing

In this chapter we have described some early activities of the FlavourCrusader initiative including interviews with "social media–savvy" foodies and the prototyping of an Australian mobile seasonal produce and recipe guide. These activities have allowed us to investigate and reflect upon the potential for mobile and social technologies in encouraging greater consumption of fresh local produce, with the intent of increasing food security, reducing emissions, improving health and supporting local agricultural communities.

The interviews highlighted diverse social, ethical, environmental, and personal motivations behind the choice to purchase local fresh produce. They also provided insight into the potential role that mobile and social technologies might play in encouraging such choices. This included the potential to leverage connections with food producers and other foodies in social networks to promote local produce, as well as ways to enable more-informed food choices at point of purchase. However, developing tools for an audience driven by ethical and social concerns limits their reach and impact. Inspired by the desire to achieve broader behavior change around fresh-food consumption, a focus on seasonal produce allowed us to promote local produce consumption to those motivated by aspects of cost, flavor, and quality, and moved the focus beyond farmers' markets. It also allowed us to respond to some of the current barriers to "buying local" identified through the interviews, such as limited access to farmers' markets and insufficient knowledge about seasonality.

Testing the prototype demonstrated the potential of a mobile tool to support different aspects of meal planning and point-of-purchase decision making. However, it also showed that providing information about seasonality alone was insufficient to support behavior change. The pathways that emerged from the test scenarios suggested that meal planning most often began with a theme or recipe, rather than an individual ingredient. To be effective, the promotion of and information about seasonal or local produce needs to be integrated into the meal-planning process, through seasonal recipes for example. A focus on meal planning and recipes could also encourage the use of unfamiliar seasonal produce by helping people understand during the shopping process how such items could be used in a meal.

Testing also reinforced opportunities for geo-targeting to increase the convenience of purchasing seasonal and local produce by displaying merchants relevant to the user's current location or target destination (e.g., near home). There was also an opportunity to actively guide people toward purchasing local produce through alternative channels by increasing their awareness of independent merchants and fresh produce markets as they searched for potential locations to purchase items. People may be further prompted to buy from such merchants if trusted peers provided recommendations.

In addition to supporting the inclusion of seasonal produce in meal planning, our research suggested ways that contextually relevant promotions from local merchants could be delivered via mobile and social tools. Such prompts could trigger consideration of fresh local produce, in turn increasing purchases.

While our initial activities have suggested opportunities for utilizing mobile and social technologies to support and encourage the consumption of local produce, it has also highlighted challenges. Developing and maintaining recipe and merchant data, for example, would present serious resource issues for a voluntary initiative such as FlavourCrusader and in some ways would duplicate the efforts of many existing recipe and food sites. The workload of community management for any crowdsourcing or recommendation system is equally daunting. An advantage of the nature of social networks and their related platforms is that rather than develop these from the ground up, we can utilize APIs and other "feeds" to link into existing data sources and communities. Mining existing recipe sites for those that fit our seasonal profile is one example. Key next steps for the project include identifying how existing social and mobile technologies that relate to meal planning can be leveraged in the name of encouraging local produce consumption.

Whether our choices about fresh food are informed ones or based on convenience, they are deeply personal and embedded into our day-to-day lives. Social and mobile tools now occupy a similarly personal space offering opportunities to access and influence our relationship to food and what we eat. In order for these tools to be leveraged for behavior change, however, we need to go beyond just the provision of information through such channels. To influence purchase decisions we need to understand how fresh-food choices are integrated into broader considerations about what to eat and cook, and attempt to support and encourage new purchase behaviors within people's meal planning process.

Acknowledgments

For their support in the activities represented in this chapter we would like to thank Sharon Lee, Social Innovation Sydney, Edmund Horan at RMIT University, and all the participants in our interviews and testing workshops.

ORECCHIETTE CON BROCCOLI

Makes about 4 servings

This dish originates from the region of Puglia in Italy. The cut of pasta is called "orecchiette" or "little ears" to denote its shape. Traditionally the recipe calls for rapini, or broccoli rabe, a slightly bitter leafy green.

The recipe is more than the sum of its parts. The sauce, made from anchovy, oil, garlic, and red chili, will pool inside the orecchiette. So a forkful is salty, hot, and full of fresh broccoli *all at once*—it's a winner!

FlavourCrusader founder Sharon Lee shared this recipe with a friend on Twitter who asked for vegetable-based recipes. He tweeted that he has made it—more than six times since. It's a great winter warmer that's easy enough for a weekday dinner.

Ingredients

1 lb. orecchiette pasta

½ lb. broccoli

4 tbsp. olive oil

2 cloves of garlic, finely chopped

1 red chili

3 anchovy fillets packed in oil

Salt and freshly ground black pepper

Preparation

1. Peel the stem of the broccoli. Discard the woody part of the stem. Cut the florets and stems into 2-inch pieces.
2. Bring a large pot of water to a boil. Add the broccoli; blanch for 5 minutes or until barely tender. Remove the broccoli with a slotted spoon and keep warm. Reserve the water.
3. Add salt to the broccoli water and bring it back to a boil. Drop in the pasta, stir well, and cook until it is al dente.
4. While the pasta is boiling, heat the olive oil in a small pan. Add the garlic, chili, and anchovy fillets. Crush the anchovy into the oil with the back of a fork. Cook over very low heat for 3–4 minutes. Add a few tablespoons of the broccoli water.
5. Drain the pasta and return it to the pan. Add the broccoli, the anchovy sauce, and black pepper to taste.
6. Serve immediately.

Notes

1. Participants were recruited using the criteria of being regular shoppers at farmers' markets or purchasing local fresh produce from other outlets. All nonindustry participants were active users of either Facebook or Twitter.

2. Some participants noted that this may not be actually the case, but that the trust engendered by closer relationship with producers fed into this perception.

3. We acknowledge/recognize the problematic relationship of "food miles" to agricultural sector emissions (see, for example, Weber and Matthews 2008) but note that food miles was a metric of which research participants were particularly cognizant.

4. This lack of trust was in part attributable to the prominence of certified organic products in major grocery retailers' promotional activities and "home brand" offerings, with distrust of retailers effectively spilling over to certification schemes.

5. The Django (https://www.djangoproject.com) and jQTouch (http://jqtjs.com) frameworks were employed for the prototype development.

6. While we acknowledge that this scenario is not strictly a "mobile" scenario, as Google (Singel 2011) notes, 81 per cent of Australians use their smartphones at home. As such we considered it an appropriate scenario for our tests.

7. Data sourced from the Farmers' Market Association of Australia (http://farmersmarkets.org.au/markets).

8. "Crowdsourcing" refers to the employment of labor or information contributed by the general public to a project, often without financial compensation.

9. "Astroturfing" is a term given to the deceptive practice of presenting an orchestrated marketing or public relations campaign in the guise of unsolicited comments from members of the public.

10. Participants suggested they would compare prices on individual produce items, outside of the context of a store (for example, comparing the price of broccoli at multiple stores while on the bus). We hypothesize that such considerations are more likely to apply at the merchant/store level, rather than individual produce item level.

11. While the prototype application was not optimized for performance, it is our opinion that such tuning would not sufficiently improve the load time of the site to support context of use.

References

Apple Inc. 2011. Apple—iOS 5 2011. http://www.apple.com/au/ios.

Australian Farmers' Market Association. 2010. Markets directory. http://farmersmarkets.org.au/markets.

Australian Social Innovation Exchange. 2011. Be part of the first Social Innovation Camp 2010. http://www.asix.org.au/sicamp.

Beanhunter. 2011. Cafe & Coffee Reviews, Coffee Online Guide, Cafe & Coffee Blog. http://www.beanhunter.com.

Brand Story and Horticulture Australia Limited. 2010. *Delivering to consumer needs: Ways to help consumers overcome barriers to vegetable purchase and consumption.* AUSVEG. http://ausveg.com.au/LiteratureRetrieve.aspx?ID=62413.

Condé Nast Digital. 2011. Epicurious.com: Recipes, Menus, Cooking Articles & Food Guides 2011. http://www.epicurious.com.

Cooler Solutions Inc. 2011. *Design for change: Eating sustainably.* http://coolersolutionsinc.com/change/wp-content/uploads/2011/07/FoodSustainability2011_Jul12.pdf.

Coster, Max, and Nicole Kennon. 2005. "New Generation" Farmers' Markets in Rural Communities. Rural Industries Research and Development Corporation. https://rirdc.infoservices.com.au/downloads/05-109.

Department of Agriculture, Fisheries, and Forestry (Commonwealth of Australia). 2006. *Innovation: The future of agribusiness* 3 (6).

Fogg, B. J. 2009. A behavior model for persuasive design. In *Proceedings of the 4th International Conference on Persuasive Technology*. Claremont, CA: ACM.

Gaballa, Sophie, and Asha Bee Abraham. 2008. *Food miles in Australia: A preliminary study of Melbourne, Victoria*. CERES Community Environment Part. http://www.greenjourney.com.au/attachments/156_Food Miles in Australia.pdf.

Hagen, Penny, Toni Robertson, Melanie Kan, and Kirsten Sadler. 2005. Emerging research methods for understanding mobile technology use. In *Proceedings of the 17th Australia Conference on Computer-Human Interaction: Citizens Online: Considerations for Today and the Future*. Canberra, Australia: Computer-Human Interaction Special Interest Group (CHISIG) of Australia.

Hendy, Nina. 2010. How to get a slice of Australia's $1 billion organic market. *SmartCompany*. http://www.smartcompany.com.au/sales/20100923-how-to-get-a-slice-of-australia-s-1-billion -organic-market.html.

Hogan, Lindsay, and Sally Thorpe. 2009. *Issues in food miles and carbon labelling*. ABARE research report 09.18, Canberra, December. http://adl.brs.gov.au/data/warehouse/pe_abarebrs99001677/foodmiles.pdf.

Jamie Oliver Enterprises. 2011. Mobile apps. http://www.jamieoliver.com/apps.

Kjeldskov, Jesper, and Mikael B. Skov. 2007. Studying usability in sitro: Simulating real world phenomena in controlled environments. *International Journal of Human-Computer Interaction* 22:1–2, 7–36.

Lee, Sharon, and Michelle Williams. 2011. Help create the future of food. *Social Innovations Sydney 2011*. http://socialinnovationsydney.org/2011/02/help-create-the-future-of-food.

Lockie, Stewart. 2008. Responsibility and agency within alternative food networks: assembling the "citizen consumer." *Agriculture and Human Values* 26:193–201.

Manzini, Ezio. 2007. Design research for sustainable social innovation. In *Design research now*, ed. Ralph Michel. Basel: Birkhäuser.

McKinna et al. 2007. Marketing plan for the Tasmanian vegetable industry. Department of Primary Industries, Tasmania. http://www.stors.tas.gov.au/item/stors/b0504b67-8f1f-d70c-512a -2065d38df719/1/web2/marketingplanforthetasmanianvegetableindustrynov07/images/veg%20 plan_vol%201%20marketing%20plan%20summary.pdf.

Meat and Livestock Australia Limited. 2008. The real food trend. http://www .fedupwithfoodadditives.info/features/consumers/The%20Real%20Food%20trend.pdf.

Moses, Asher. 2011. Australia's white hot smartphone revolution. *Sydney Morning Herald*, September 8.

NM Incite. 2011. *State of the media: The social media report*. Nielson. http://www.nielsen.com/content/dam/corporate/us/en/reports-downloads/2011-Reports/nielsen-social-media-report.pdf.

OpenIDEO. 2011. How might we better connect food production and consumption? *OpenIDEO*. http://www.openideo.com/open/localfood/inspiration.

PricewaterhouseCoopers. 2007. The economic contribution of small to medium-sized grocery retailers to the Australian economy, with a particular focus on Western Australia. The Productivity Commission, Australia. http://www.pc.gov.au/__data/assets/pdf_file/0003/66540/sub082.pdf.

Priebatsch, Seth. 2010. *Building the game layer on top of the world*. Boston: TED. Video.

Robinson, Les. 2011. Five doors: An integrated theory of behaviour change. *Enabling change*. http://www.enablingchange.com.au.

Schonfeld, Erick. 2010. SCVNGR's secret game mechanics playdeck. *TechCrunch*. http://techcrunch.com/2010/08/25/scvngr-game-mechanics.

Singel, Ryan. 2011. Google recipe search cooks up next gen of search. *Wired*. http://www.wired.com/epicenter/2011/02/google-recipe-semantic.

Social Innovation Sydney. 2011. Social Innovation Sydney 2011. http://socialinnovationsydney.org.

The Monday Campaigns Inc. 2011. Meatless Monday—One day a week, cut out meat 2011. http://www.meatlessmonday.com.

Urbanspoon. 2011. Urbanspoon. http://www.urbanspoon.com.

Vermeir, Iris, and Wim Verbeke. 2006. Sustainable Food Consumption: Sustainable food consumption: Exploring the consumer "attitude–behavioural intention" gap. *Journal of Agricultural & Environmental Ethics* 19 (2):169–194.

Weber, Christopher L., and H. Scott Matthews. 2008. Food-miles and the relative climate impacts of food choices in the United States. *Environmental Science & Technology* 42 (10):3508–3513.

What Is It Production Ltd. 2011. Seasons—For the iPhone 2010. http://www.seasonsapp.com.

Wikipedia. 2012. BarCamp 2012. http://en.wikipedia.org/wiki/BarCamp.

Winter, Michael. 2003. Embeddedness, the new food economy and defensive localism. *Journal of Rural Studies* 19 (1):23–32.

Young, Grant. 2010a. FlavourCrusader Personas. http://synapsechronicles.com/wp-content/uploads/2011/02/flavourcrusader-personas.pdf.

Young, Grant. 2010b. Report on design research with urban local food customers. http://synapsechronicles.com/wp-content/uploads/2011/02/design-research-with-urban-local-food-customers.pdf.

8 Probing the Market: Using Cultural Probes to Inform Design for Sustainable Food Practices at a Farmers' Market

Eric P. S. Baumer, Megan Halpern, Vera Khovanskaya, and Geri K. Gay

Environmental sustainability has become an increasingly pressing issue, gaining significant attention from popular media, government agencies, and academic researchers in a variety of disciplines. Some such work involves reducing harmful emissions, which result from transporting goods, by purchasing goods closer to their point of origin, that is, buying locally. While not always an option for every type of product, locally grown food, especially fresh produce, is a viable alternative for many to mass-produced or processed foods. Farmers' markets, periodic gatherings where farmers come to sell their crops directly to consumers, represent one means of buying food locally. Previous work has examined the cultures and practices of farmers' markets and how they integrate with their broader community (Alkon 2008; Brown 2002; Lyson, Gillespie, and Hilchey 1995; McGrath, Sherry, and Heisley 1993; Slocum, Ellsworth, Zerbib, and Saldanha 2009; Trobe 2001).

The field of human-computer interaction (HCI) has also experienced an influx of research on the potential relationships between information and communication technologies (ICTs) and environmental sustainability (Blevis 2007; DiSalvo, Sengers, and Brynjarsdóttir 2010; Froehlich et al. 2009; Tomlinson 2010; Woodruff, Hasbrouck, and Augustin 2008). For example, some have argued that food production, distribution, and purchasing are environments where ICTs may be beneficially deployed to improve and enhance environmentally sustainable practices (e.g., Hirsch, Sengers, Blevis, Beckwith, and Parikh 2010). One project, though not specifically sustainability focused, developed an augmented reality system for farmers' market visitors (Light, Wakeman, Robinson, Basu, and Chalmers 2010). Such work points not only to opportunities but also to complexities in this area, such as the ways in which sustainability is defined, constituted, and enacted in local, culturally specific ways (e.g., DiSalvo, Sengers, and Brynjarsdóttir 2010).

Many of these same complexities emerge in the case of farmers' markets. Shopping at the farmers' market may serve such varied ends as living in a more environmentally sustainable manner, investing in local economic development, supporting fair labor practices, acquiring the freshest and tastiest produce, or "seeing and being seen"

(Alkon 2008; Lyson, Gillespie, and Hilchey 1995; McGrath, Sherry, and Heisley 1993; Trobe 2001). Furthermore, these motivations are not mutually exclusive; a visitor to the market seeking the freshest kale available may simultaneously want to invest in the local agricultural industry. While these complexities and interconnections make the farmers' market a rich site for study, understanding them introduces methodological challenges.

In this chapter, we argue that cultural probes (Boehner et al. 2007; Gaver et al. 2004; Gaver, Dunne, and Pacenti 1999) can be an effective means of exploring design possibilities for sites of sustainable food practices where various interconnected concerns may be at work. We draw on our experiences using cultural probes at the Ithaca Farmers' Market in Ithaca, New York, to describe the ways in which cultural probes were and, at times, were not useful in terms of providing a rich picture of visitors' experiences at the market to help inform design. This chapter's contribution, then, is primarily methodological, in terms of providing practical understanding and guidance for using cultural probes in such settings. However, our process of interpreting participants' responses raised a number of larger questions in terms of designing technology around food and sustainability. Thus, this chapter also makes a conceptual contribution by providing an argument for how we as researchers might go about understanding, engaging with, and designing (or not designing) for sustainability in the context of local food practices and cultures.

Farmers' Markets

As mentioned above, much previous work on farmers' markets has focused either on the economic or on the sociocultural aspects thereof, and sometimes the relations between the two. For example, Alison Alkon (2008) compared two farmers markets— one in Berkeley, California, and one in Oakland, California—highlighting differences in terms of primary motivations. While buying locally and other environmental concerns were forefront at the Berkeley market, social justice and equal access to healthy, nutritious food predominated at the Oakland market. The work of Rachel Slocum and colleagues (2009) involves similar concerns, focusing on the racial make-up of visitors to a market in Minneapolis, Minnesota. Their findings suggest significant racial differences in shopping practices; the importance of buying locally grown produce, the amount of money spent, and the importance of avoiding products grown with pesticides, growth hormones, or other non-organic techniques all varied among different racial and ethnic groups. Lidia Marte (2007) also examines the relationship between race and food, but rather than focusing on any one farmers' market, she takes a broader view, developing a technique she calls food mapping. This approach involves asking people to depict—through photographs, sketches, and interviews—not only where food physically comes from, but also both the distinct personal and broader sociocultural narratives that give rise to, and are enacted through, cooking and eating. In their ethnography of an American Midwest farmers market, Mary Ann McGrath, John F. Sherry, and Deborah D. Heisley (1993) describe the different types of people who attend the market, from early birds, to socializers, to discount hunters, and how they each contribute differently to the market atmosphere. In surveying visitors to a farmers' market in the United Kingdom, Helen L. Trobe (2001) found that motivations often included avoiding genetically modified organisms in favor of organically grown produce, and that such organic foods were perceived to be both healthier and more flavorful.

While this work collectively provides a useful understanding of the social and economic landscape of farmers' markets, less work has been done considering how one might design for this space. In one exception, Ann Light and colleagues (2010) developed an augmented reality application that enabled market visitors to annotate the physical space of the market with information about vendors or products, as well as with stories about their own experiences of the market. While the system elicited interesting information from users, it also pointed to ways in which such technology can interfere with traditional vendor-visitor interactions. These various studies both provide an intellectual context for this work and suggest a number of pressure points that may be worth exploring.

Cultural Probes

Cultural probes is a design technique for exploring—or "probing"—a specific cultural setting. Study participants complete a variety of open-ended, provocative, and at times oblique or ambiguous activities. The probes themselves are highly designed, sometimes personalized objects, and they are intended to facilitate a conversation between a designer and members of the community for which she or he is designing.

William W. Gaver, Tony Dunne, and Elena Pacenti (1999) developed cultural probes to explore different means of integrating elderly residents into the neighborhoods in which they live. Physically, the probes themselves were packets containing materials for completing a variety of activities. For example, the packets included several postcards, each of which included a reflective prompt, such as "Tell us about your favorite device." Postcards were used to guide both the length and style of participants' responses. The packets also included disposable cameras with instructions to photograph such things as "your home," "the first person you see today," "something desirable," and "something boring." Maps of the neighborhoods were also included, along with instructions to indicate such places as "where you go to be alone," "where you like to day dream," and "where you want to go but can't." Cultural probes provide a window onto the complexities that lie beneath the surface of everyday community encounters in order to design artifacts that are relevant to that community.

Here it is important to note that cultural probes are not a means of formal requirements gathering, of obtaining implications for design, or of conducting "ethnography by post" (Boehner et al. 2007). They do not generate data that are then analyzed but rather provide "inspirations for design" (Gaver, Dunne, and Pacenti 1999). To be sure, cultural probes share ethnography's emphasis on interpretation, but this emphasis plays out differently. First, the value of cultural probes is in participants' interpretations of the prompts and activities. Often times, instructions are intentionally vague or ambiguous, such as the postcard and map prompts. The most effective probes are not those that evidence patterns across responses but those that provoke diverse responses, which can be illustrative in understanding participants' unique cultural context. Second, as mentioned above, these responses require interpretation by the designers or researchers conducting the probes in order to be made meaningful, and the same response can often be interpreted in multiple ways. While a researcher's interpretation is an important component of any study, the ambiguity of the probes and the responses they elicit foreground these processes of interpretation. Thus, while the activities in which study volunteers engage may resemble some ethnographic techniques, such as Marte's (2007), responses to these activities are treated in a different manner.

Probes must also be highly designed objects. They are meant to provoke aesthetic experiences, which we interpret as similar to those in John Dewey's (1934) description of art as experience. Dewey describes a constant state of experiencing our surroundings; however, he reserves the distinction of an experience for those points when "the material experienced runs its course to fulfillment" (1934, 37). For an experience to occur, Dewey says that there can be "no division between act and material, subject and object" (Dewey 1934, 10–11; McCarthy & Wright 2004, 54) and that this lack of division is partially constitutive of an experience.

An example of such attention to aesthetic experience can be seen in the work of Gaver, Dunne, and Pacenti (1999), described above. The disposable cameras given to participants had the original manufacturer's standard packaging removed and replaced with instructions for how to use the camera and prompts for taking photos. The goal was to make the camera, and the probes in general, "aesthetically crafted, [but] not too professionally finished" (Gaver, Dunne, and Pacenti 1999, 26). A disposable camera with Kodak packaging would take participants out of the experience of the probe by placing the easily recognizable logo, marketing language, and cautions found on such labels between the "self and object."

To provide users with probe materials that were not developed to evoke the earthiness or hand-made-ness of a farmers' market would have similarly impeded participants' experiences. Furthermore, we wanted to design the probes in such a way that they not only provided insights into the meaningfulness of participants' practices at the market, but also allowed participants to gain something from the time they spent completing the activities. Focusing on providing our participants with aesthetic experiences provided an effective way of addressing both these concerns.

Designing a Diary

For several reasons, we chose to shape our farmers' market probe as a week-long diary, where each day asked the participant to complete a different activity. First, the diary framing was intended to elicit close, intimate impressions and experiences of the market. Second, the farmers' market we studied in winter occurs once a week, so making the diary one-week long allowed it to coincide with the market's regular cycle.[1] Third, the diary format provided a convenient means of organizing varied activities and prompts under a single conceptually coherent umbrella.

The physical diaries themselves were printed on recycled paper and stitched together into a bound booklet, the covers of which were each painted with a different color gouache (figure 8.1, left). These colored covers not only provided a custom and hand-made feel, but the gouache also added a unique tactile element. The first page of the diary included a table of contents (figure 8.1, right), followed by one activity per page. Each activity page followed a dominant color theme and included a title, the

Figure 8.1
Cover and contents pages from the diary

appropriate day of the week, a prompt, and a key image, which resembled a stylized water color. All images were hand-painted using gouache and then scanned into a computer, providing for "aesthetic control" (Gaver, Dunne, and Pacenti 1999, 24) and facilitating an aesthetic experience, as described above. The remainder of this section describes some of the activities and their accompanying design rationale.

Reflecting

Participants were asked to complete the first day of their diary upon returning home from the market. The prompt asked, "Tell us about your trip to the market today. How did it go? Whom did you see? Did you have any good finds? Was there anything disappointing?" We wanted the prompt to be somewhat open-ended, allowing participants to tell us about any aspect of their trip, but still provide guidance as to the types of things that might be of interest to us.

Mapping

On the second day, participants were asked to draw a map of the market, a common technique in related work (Lynch 1960; Marte 2007; Gaver, Dunne, and Pacenti 1999). On a fold-out sheet of paper sewn into the binding, the diary asked participants to draw a map of the market and indicate on that map "your favorite vendor, the busiest spot, the least-local products, a lonely place, vendors you trust, [and] the tastiest produce." This page used a rustic, off-center compass rose as its key image to reinforce that the map did not need to represent the market perfectly accurately (figure 8.2A, left). These indicators were chosen to provide participants specific direction while simultaneously allowing us to see how participants interpreted, for example, what "least local" or "tastiest" might mean.

Drawing

In designing the diary, we sought to include a variety of activities and prompts. Day 4 asked participants, "Draw a picture of the people you go with or see at the market and their favorite thing from the market." This prompt was accompanied by an abstract "bottomground" of green grass above which participants could draw other visitors to the market. We hoped this activity would help us glimpse not only the social and community-oriented experience of the market, but also how these aspects connect to other things at the market.

Budgeting

The cost of produce at farmers' markets (Shah 2010) and the amount of money an individual or family spends (Slocum et al. 2009) can both be complex potentially divisive topics. We wanted to understand how market visitors negotiate with themselves how to spend their money at the market. The Budgeting activity instructed

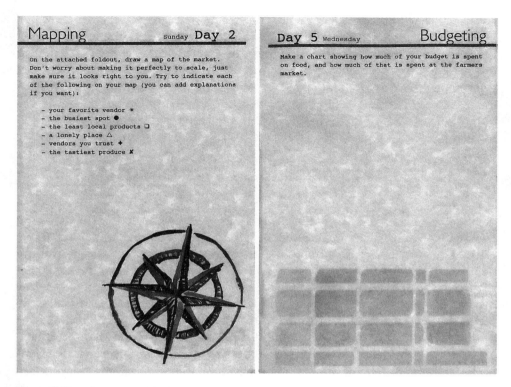

Figure 8.2
The second and fifth pages from the diary

participants to "Make a chart showing how much of your budget is spent on food, and how much of that is spent at the farmers market." This prompt was accompanied by an abstract grid at the bottom of the page designed to be reminiscent of an accounting register (figure 8.2, right). The grid, however, was somewhat irregular, since accounting for personal finances can often be somewhat irregular.

Preparing
The final day of the diary asked participants, "How are you preparing for your next trip to the market? Are you making a list? Is there anything in particular that you're looking for?" This activity was intended to help illuminate the part of the market routine that happens before or leading up to the market visit. It was also intended to be a companion to the Reflecting activity on the first day of the diary; Reflecting looks backward at the previous visit, while Preparing looks forward to the next visit.

Across all these activities, there were a number of other steps taken to ensure the wholeness of participants' aesthetic experiences with the diary. Each day's activity was given a title in the form of a present progressive verb (e.g., Mapping, Drawing), intended to connote being active and engaged. The hand-made aesthetic of the object was meant to help participants feel that they could put sketches or vague ideas in their diaries rather than polished content. In writing the prompts, we strove to keep use of first-person pronouns to a minimum. The contents page mentioned "we," and Reflecting asked participants to "tell us" about their trip to the market, but these were the only instances of the first person. The goal was to keep the inside of the diary as personal as possible by keeping the researchers as absent as possible, thereby encouraging diary-like use of the probe.

Distributing the Diaries

We found people to complete our diaries in a variety of ways, including emails to local environment- and agriculture-oriented list serves, postings on similar websites, physical flyers at the market itself and on community bulletin boards, and by word of mouth. After a participant contacted us, one of us met her or him in person, either at the farmers' market itself or at a mutually convenient place and time, to deliver the diary and answer any questions. Once participants had completed the diary, either we met them again to collect it, or participants returned it to us via postal mail.

Although it was a driving motivation for the study, explicit mention of sustainability was kept to a minimum—throughout the recruitment materials, our directions about how to complete the diary, and the diary itself. The only activity that included express mention of environmental concerns was Reacting, in which participants responded to two provocative quotes, but the quotes were more about farming and agriculture than about sustainability per se. Rather than focusing exclusively on sustainability, we wanted to see if it would emerge on its own. Thus, we sought to understand the experiences of the market in a holistic way, both the many varied aspects and the connections between them.

In total, we gave out twelve diaries. Eight of these were returned to us; two participants did not make it to the market before the end of the winter market season, and two participants stopped responding to emails about returning the diaries. These diaries were returned over the course of about a month, from early February to early March 2011.

Once all the diaries were returned, this chapter's first author, Eric Baumer, read through each of them, taking notes about interesting, compelling, or confusing responses. Then, the Baumer and chapter co-authors Megan Halpern and Vera Khovanskaya read the diaries together. All the completed diaries were laid out on a table, and each was opened to the first page, so that all the responses to the first activity could be read side-by-side and discussed. Each set of responses was read and discussed in a similar manner. From these conversations emerged a number of design ideas, but not exactly of the type we were expecting.

Interpreting the Responses

As described above, cultural probes are intended not as a means of data collection but as a means of inspiring design. While reading through the diaries, it became clear what worked well in terms of eliciting varied and provocative design inspirations from participants, as well as which prompts were less effective. In fact, a significant portion of our discussion about participants' responses to the diary had to do with the design of the diary itself. This section, then, focuses on the methodological contribution of this paper, providing practical guidance for the design of such cultural probes based on our experiences.

The primary insight is that, while we thought them open-ended, many of our prompts were ultimately too prescriptive. For example, when we wrote the Reflecting prompt, we intended to give participants a sense of the kinds of questions they might consider in reflecting on their trip to the market. However, many participants simply responded to it as a series of questions rather than engaging in any sort of deep reflection about their trip to the market. Responses to the Eating activity, which asked participants to draw their favorite meal that week and indicate which portions came from the market, followed this pattern. One participant, stating that she did not draw well, cut out a picture from a magazine showing yogurt with granola and berries. She described how she had purchased the berries at the market the previous summer and then frozen them. Other responses to this prompt included other detailed and varied descriptions of what people ate, but they were less effective at helping us understand what eating that food meant to them. Similarly, in the Preparing activity, participants described what they do in the market-preparation routine, but not why they do it. In general, we received numerous detailed accounts of peoples' experiences around and related to the market, but not the interpretations or the meanings of those experiences.

One prompt that worked surprisingly well was Budgeting, generating highly varied responses. Most people presented their budget in monthly terms, with anywhere from less than one-fifth to more than one-half their money being spent on food. The portion spent at the market was anywhere from just a sliver to well over half of their food budget. People's financial situations also varied, ranging from people who do not know how much they spend on food because they have not needed to budget in years, to those whose trips to the market are constrained by living on a fixed income or social security disbursement.

What was most interesting was not necessarily what participants said about spending their money but how they said it. The ways in which people presented their budgets were incredibly varied, including pie charts, subdivided tables, colored graphs, and flow charts. The phrasing was also at times highly informative. In a paragraph explaining her budget chart, one participant described how she and her partner "live a radically simple lifestyle with zero debt (no mortgage, no credit cards, no car payments)." There are numerous different ways in which one's life could be simple or complex; why is financial simplicity so dominant as to enable "radical" simplicity? We found these responses particularly interesting because, while designing the diaries, we were concerned that the budgeting activity might end up being too boring. Instead, it tapped into something that held significant meaning for the participants, and their responses facilitated reflection (both theirs and ours) on this significance. These results also provide an interesting complement to the work of Slocum and colleagues (2009). While their results show informative differences in how different groups of people spend money at the farmers' market they studied, our findings shed light on why people spend money the way they do and the reasoning processes behind those spending decisions.

A Diary Redesigned

As this summary suggests, the diaries were very effective at eliciting the practical details of interactions with and around the farmers' market, but less so the experiential details. Thus, we redesigned the diary, revisiting each activity to make it less prescriptive, more open-ended, slightly more oblique, and, hopefully, more effective at eliciting provocative responses and interpretations. In short, we sought to create more opportunities for the diary to facilitate aesthetic experiences.

For example, the Reflecting prompt was changed to say simply, "The market today was . . ." (an approach we hoped would generate more varied responses than the previous question-laden version). The Eating activity was renamed "Meaning," and the prompt to "Draw what food from the farmers' market means to you." The intent was to generate responses that help us understand the meaningful ways in which the market connects to the rest of participants' lives. The final Preparing activity was replaced entirely with an activity instead called Completing, the prompt for which read, "What is missing from this diary?" Inspiration for this prompt was taken from qualitative interview work done by Eric Baumer and Megan Halpern, in which each would often end interviews with the following two questions: "What other questions have I not asked you?" and "Do you have any questions for me?" We wanted to capture the spirit of these questions, but also maintain the aforementioned absence of the researcher from the diary by eschewing first-person pronouns. Furthermore, we believed that this prompt should still serve as a companion to the opening Reflecting prompt, but this time encourage participants to reflect not on their experiences of the market but rather on the lens that the diary provides for those experiences.

These redesigned diaries were distributed to eight more participants, six of whom returned them. Overall, responses differed in three specific ways. First, they were usually shorter and less detail oriented. For example, one response to Reflecting said the market was "lively & vibrant, even though it was close to the market shutting

Figure 8.3
Example responses from the second round of diaries

down for the day," and another simply had the words "Fun," "Aromatic," "Tasty," and "Colorful" arranged on the page at various angles. In another example, when asked to draw what food from the market means, one participant simply drew a bunch of greens with the caption "just real food" (figure 8.3, left).

Second, most responses, although shorter, were also more provocative—sometimes due, in part, to their conciseness—and gave a better impression of the subjective experiences of the market. In addition to the examples above, one Reflecting response described the market as "crowded, but slow / sunny, breezy, warm, lovely / . . . calming / intermediate, before the rest of the day." Another diary's response to the Completing prompt mentioned "Other farmers markets that I visit: DeWitt Park, Lansing Farmers Market." These and other responses brought up aspects of the market experience we had not anticipated.

Third, reading through the diaries, each one felt more coherent. The responses to different prompts fit with one another, often times evidencing the same themes. For example, one response to Reflecting mentioned that she "got there [to the market] early, so it wasn't overwhelming stress-wise." In describing her map, she mentioned that she tries "to beeline to [her] tried & true sources of food. It's all a blur of stress." For the Experiencing prompt, she drew a depiction of herself feeling stressed out (figure 8.3, right). Another participant, in response to the Experiencing prompt, described the diary itself as "a bit cutesy & therapy-derived." For the final prompt, she wrote, "I love the place [the farmers' market], I'm there every weekend it's open & I'm in town, but it's not for romantic reasons!" Thus, each diary helped construct a coherent picture of a single individual's experiences of the market. Thus, in these ways, the redesigned diaries were rather more successful. This is not to say that the first round was useless. On the contrary, iterating on their first round, and comparing responses between the first and second round, helped us gain a richer understanding of how people experience and interact with the market.

Discussion and Methodological Considerations

What the participants' responses in the diaries tell us about their experiences of and around the market extend beyond the scope of the chapter. Instead, we address here a few methodological and conceptual issues raised in this work. First, our choice to use diaries contrasts with most previous work on farmers' markets, which usually employs ethnographic methods, such as open-ended interviews and participant observation,[2] and one might wonder why we didn't use such methods here. Part of the value of using a diary for a study such as this is that it physically accompanies the participant to her or his home. She may read the prompt and respond immediately, or she may leave the diary, think for a while, do something else, and return to respond later. This is not to say that interviews would not be valuable. Indeed, in follow-up work, we are conducting interviews and focus groups with participants who completed these diaries. However, it is important to note that such interviews are intended to complement the diaries. While they may provide further insight, we do not believe that interviews are a prerequisite for using diary responses to inform and inspire design (cf. Gaver, Dunne, and Pacenti 1999).

Second, it is tempting to distill from the diaries general findings about different types of farmers' market experiences. However, to reiterate, the goal here is not to provide findings that might resemble an ethnographic account of the farmers' market or associated food practices; many exemplary cases of such work exist (e.g., Alkon 2008; Lyson, Gillespie, Hilchey 1995; McGrath, Sherry, and Heisley 1993; Slocum et al. 2009; Trobe 2001). Rather, the goal is to provide insight and inspiration as to how one might design for sustainability within and around the space of a farmers' market. The description of our participants' responses presented here provides some such insight; the subsequent section describes specific design inspirations that arose from these responses.

Finally, it is important to clarify how this work pertains to sustainability. We argued in the beginning of this chapter that, while sustainability is an important aspect of the farmers' market, the market is about more than just sustainability. The diaries were useful not only at drawing out concerns and experiences relevant to different aspects of the market, but also at allowing participants to draw connections and contrasts among those aspects. For example, responses to the Reflecting prompt included visual, tactile, and olfactory, as well as social and personal aspects; responses to prompts about what the farmers' market means often included economic, social, environmental, health, and other concerns. Rather than helping to build an understanding of the complex interconnections among sustainability and various other issues, the experiences presented in this chapter consider how designing for (sustainability at) the market might take into account such complexities.

Implications and Conversations

In our exploration of the farmers' market, the diaries generated design inspirations, but not exactly in the way expected. One set of design concepts dealt directly with the experiences of the market, but not involving technology—at least not the kinds of computational and information technologies with which we are accustomed to working. We found ourselves wanting to extend the conversations begun by the diaries to other market visitors via some sort of intervention at the market itself, perhaps with something as simple as a prompt from the diary reproduced at the top of a chalkboard or poster-sized sheet of paper. Another idea we considered involved Polaroid cameras. We could give market visitors a camera and a series of things to photograph, such as "community," "profit," "waste," or "sustainability." Participants could then pin their photographs on a poster, annotate them with markers, or draw annotations connecting multiple photos. These design interventions are interesting for two reasons.

First, they extend cultural probes to involve the community. Cultural probes were intended as a means of fostering a conversation between a designer and members of a community. What we describe here might be called community probes: interventions by a designer intended to foster conversations and shared experiences among community members. This idea goes beyond the typical designs informed by cultural probes, such as the Sloganbench (Gaver and Beaver 2006), which displayed provocative quotes gathered from community members by designers. A community probe, in contrast, would incorporate reactions and conversations around such a design intervention into the artifact itself. We suggest that future work may benefit from exploring this opportunity. Not only does it offer a chance for design research that engages the community, but it also provides a unique opportunity to consider the designer's role in the community (Manzini 2006).

Second, these results in general speak to the appropriateness of using cultural probes for contexts such as this. As mentioned previously, all the authors of this chapter are, to some extent, technology designers.[3] Had we engaged in a traditional technology-system design approach—requirements gathering, prototype building, preliminary usability testing, et cetera—we likely would have ended up designing some sort of technological system to support experiences of sustainability at the farmers' market. Such design methods are not particularly adept at handling situations such as this, when the implication is not to design (cf. Baumer and Silberman 2011), or at least not to design computational technology. The flexibility of cultural probes, however, allowed for designing interventions that were not expressly or necessarily technological. Thus, we suggest, that combined with the experiential contributions presented in this chapter, cultural probes may be a particularly useful method for exploring the design space around food, technology, and sustainability.

Acknowledgments

Thanks to Phoebe Sengers, Hrönn Brynjarsdóttir, Laura Forlano, Steven Ibara, the Interaction Design Lab (IDL), and the Culturally Embedded Computing Group (CEmCom) for helpful conversations. Thanks also to the anonymous reviews for constructive feedback and relevant citations.

PAN-GRILLED PORK CHOPS WITH STRAWBERRY RHUBARB GLAZE

Makes 2 servings
This recipe can be made (almost) entirely with ingredients available at the Ithaca Farmers' Market, the field site at the focus of our chapter. At the time of writing, soy sauce was the only ingredient not available at the farmers' market. Eric Baumer invented the recipe in early summer when the only vegetables readily available were rhubarb and radishes.

Ingredients

(All measurements were estimated post hoc, so feel free to adjust.)
1 lb. pork chops (2 chops, each roughly 1-inch thick)

2–4 stalks rhubarb (depending on length), sliced ½-inch thick, crosswise

4 oz. mushrooms, sliced

1 dozen medium-sized strawberries, sliced

6 radishes (optional), sliced

1 to 1.5 tbsp. honey

Soy sauce

Preparation

1. Sear pork chops in a grill pan over medium heat until brown on each side, but not completely cooked (meat should still yield slightly when poked with a spatula). Remove from pan keep warm. While chops are grilling, prepare the glaze.

2. In a medium-sized frying pan, sauté rhubarb in about 2 tbsp. of soy sauce for 3–5 minutes. If using radishes, sauté with rhubarb. Add mushrooms and about 1.5 tsp. of soy sauce; sauté for an additional 2–3 minutes or until mushrooms begin to change color. Add strawberries and about 1.5 tsp. soy sauce, sauté for an additional 1–2 minutes, or until strawberries become slightly mushy.

3. Move all ingredients to the sides of the frying pan, clearing a space in the middle. Place honey in frying pan and cover in about 2 tbsp. of soy sauce. Mix soy sauce and honey together for about 15 seconds, or until honey mostly dissolves. Mix with remaining contents of the pan. Place the chops on top of the fruit-vegetable mixture. Reduce heat, cover, and simmer for 8–10 minutes, flipping chops once halfway through, or until chops are done (they will be relatively firm to the touch or, when pierced with a knife, the juices run just slightly pink).

4. Plate chops, cover with fruit-vegetable-glaze mixture, and serve with couscous, rice, or fresh bread.

Notes

1. The study described here was conducted during the local winter market, which is held Saturdays during January and February. The analogous summer market runs at different times on multiple days of the week, but on the same days each week.

2. While we did not engage in ethnographic participant observation, all authors of this paper had visited and shopped at the Ithaca Farmers' Market, with varying degrees of regularity, before engaging in this line of research.

3. Each author comes to technology design from a different perspective, including computer and information sciences, theater and set design, architecture and graphic design, and communication.

References

Alkon, Alison. 2008. Paradise or pavement: The social constructions of the environment in two urban farmers' markets and their implications for environmental justice and sustainability. *Local Environment: The International Journal of Justice and Sustainability* 13 (3):271–289. doi: 10.1080/13549830701669039.

Baumer, Eric P. S., and M. Six Silberman. (2011). When the implication is not to design (technology). In *Proceedings of the 2011 Annual Conference on Human Factors in Computing Systems (CHI '11)*, 2271–2274. New York: ACM. doi:10.1145/1978942.1979275.

Blevis, Eli. (2007). Sustainable interaction design: Invention and disposal, renewal and reuse. In *Proceedings of the SIGCHI Conference on Human Human Factors in Computing Systems (CHI '07)*, 503–512. New York: ACM. doi:10.1145/1240624.1240705.

Boehner, Kirsten, Janet Vertesi, Phoebe Sengers, and Paul Dourish. (2007). How HCI interprets the probes. In *Proceedings of the SIGCHI Conference on Human Factors in Computing Systems (CHI '07)*, 1077–1086. New York: ACM.

Brown, Allison. 2002. Farmers' market research 1940–2000: An inventory and review. *American Journal of Alternative Agriculture* 17 (04):167–176. doi:10.1079/AJAA200218.

Dewey, John. 1934. *Art as experience*. New York: Perigee Books.

DiSalvo, Carl, Phoebe Sengers, and Hronn Brynjarsdóttir. (2010). Mapping the landscape of sustainable HCI. In *Proceedings of the 2010 Annual Conference on Human Factors in Computing Systems (CHI'10)*, 1975–1984. New York: ACM. doi:10.1145/1753326.1753625.

Froehlich, Jon, Tawanna Dillahunt, Predrag Klasnja, Jennifer Mankoff, Sunny Consolvo, Beverly Harrison, and James A. Landay. (2009). UbiGreen: Investigating a mobile tool for tracking and supporting green transportation habits. In *Proceedings of the 2009 Annual Conference on Human Factors in Computing Systems (CHI '09)*, 1043–1052. New York: ACM. doi:10.1145/1518701.1518861.

Gaver, William W. and Jacob Beaver. (2006). The presence project: Helping older people engage with their local communities. In Networked Neighbourhoods: The Connected Community in Context, ed. P. Purcell, 345–371. London: Springer-Verlag. doi:10.1007/1-84628-601-8_14.

Gaver, William W., Andrew Boucher, Sarah Pennington, and Brendan Walker. 2004. Cultural probes and the value of uncertainty. *interactions* 11(5): 53–56. doi:10.1145/1015530.1015555.

Gaver, William W., Tony Dunne, and Elena Pacenti. 1999. Cultural probes. *interactions* 6(1): 21–29.

Hirsch, Tad, Phoebe Sengers, Eli Blevis, Richard Beckwith, and Tapan Parikh. (2010). Making food, producing sustainability. In *Extended Abstracts of the 2010 Annual Conference on Human factors in Computing Systems (CHI '10)*, 3147–3150. New York: ACM. doi:10.1145/1753846.1753939.

Light, Ann, Ian Wakeman, Jon Robinson, Anirban Basu, and Dan Chalmers. (2010). Chutney and relish: Designing to augment the experience of shopping at a farmers' market. In *Proceedings of OZCHI 2010: Design—Interaction—Participation*, 208–215. New York: ACM.

Lynch, Kevin. 1960. *The image of the city*. Cambridge. MA: MIT Press.

Lyson, Thomas A., Gilbert W. Gillespie, and Duncan Hilchey. 1995. Farmers' markets and the local community: Bridging the formal and informal economy. *American Journal of Alternative Agriculture* 10 (03):108–113. doi:10.1017/S0889189300006251.

Manzini, Ezio. 2006. *Creative Communities, Collaborative Networks and Distributed Economies: Promising Signals for a Sustainable Development. Communities*, 1–10. Milan: INDACO, Politecnico di Milano.

Marte, Lidia. 2007. Foodmaps: Tracing boundaries of "home" through food relations. *Food and Foodways: Explorations in the History and Culture of Human Nourishment* 15 (3–4):37–41. doi:10.1080/07409710701620243.

McCarthy, John, and Peter Wright. 2004. *Technology as Experience*. Cambridge: MIT Press.

McGrath, Mary Ann, John F. Sherry, and Deborah D. Heisley. D. 1993. An ethnographic study of an urban periodic marketplace: Lessons from the Midville farmers' market. *Journal of Retailing* 69 (3):280–319. doi:10.1016/0022-4359(93)90009-8.

Shah, Riddhi. 2010, June 21. The dark side of the farmers' market boom. *Salon.com*. http://www.salon,com/2010/06/21/too_many_farmers_markets.

Slocum, Rachel, Elisabeth Ellsworth, Sandrine Zerbib, and Arun Saldanha. 2009. Local food and diversity in public space: A study of the perceptions and practices of Minneapolis farmers' market customers. CURA Reporter 39 (1–2 Spring/Summer):40–50.

Tomlinson, Bill. 2010. *Greening through IT: Information technology for environmental sustainability*. Cambridge, MA: MIT Press.

Trobe, Helen. L. 2001. Farmers' markets: Consuming local rural produce. *International Journal of Consumer Studies* 25 (3):181–192. doi:10.1046/j.1470-6431.2001.00171.x.

Woodruff, Alison, Jay Hasbrouck, and Sally Augustin. (2008). A bright green perspective on sustainable choices. In *Proceedings of the Twenty-Sixth Annual SIGCHI Conference on Human Factors in Computing Systems (CHI '08)*, 313–322. New York: ACM. doi:10.1145/1357054.1357109.

9 Re-placing Food: Place, Embeddedness, and Local Food

Katharine S. Willis, Katharina Frosch, and Mirjam Struppek

We live in a digital, networked society where immaterial transactions frame many aspects of our everyday life. But in an Internet of Things, our products and artifacts still have material presence: they have to be somewhere; they need to be physically displaced to arrive somewhere else; they cannot be easily duplicated; and they have a life cycle. All things that we consume need to travel from the place they are produced to the place in which they will be consumed. This process may cover a simple few feet or may involve thousands of miles and many weeks. In our current urbanized and globalized marketplace, food is often transported an extremely long distance from its origin to the place where it is sold. There has been some attempt to quantify the way in which food is part of a global marketplace through the introduction of the concept of "food miles" (Paxton 1994). This measures the distance food is transported from the time of its production until it reaches the market. The sheer distance for many items commonly available in supermarkets in the developed world underlines the disembedded nature of a globalized food system that creates a "placeless foodscape" (Ilbery and Kneafsey 2000, 319). As Ulrich Oltersdorf and Kurt Gedrich point out, "the place in which the food is produced and the place in which a person lives constructs less and less the framework for the food someone consumes" (2001, 66).

A growing resistance to this globalized food marketplace is seen as one of the reasons for the rise in interest in locally sourced food. This trend advocates closer connections between producers and consumers both in terms of spatial distance and personal networks. Such direct agricultural markets privilege locality and seasonality over distance and durability (Friedmann 1993). However, there is also a growing movement to take this a step further and to reconnect the person with food at the most basic level. In this way "localism provides a defensive position against the disempowering and homogenizing effects of globalisation" (Allen 2004, 169). Yet, this is complicated by the fact that there are many complex and conflicting meanings tied up in the discourse of the "local" with respect to food (Allen 2004). According to Leslie Duram and Lydia Oberholtzer, "although local foods are increasingly gaining the public's attention, there is no clear geographical delineation for 'local'" (2010, 2). Indeed, ideas of what constitutes local are both culturally and geographically situated. In the United States a study found that geographical topologies helped to shape ideas of what was local, such that urban density on the East Coast may mean that "local" in Washington, DC, is defined as being within a hundred miles from the city, whereas in Iowa "local" food is defined as that which has been made or produced within the state boundary (Duram and Oberholtzer 2010, 2). In Germany, local food is generally accepted as food produced in the "region," which is either a county or a topographic region. In a survey, 40 percent understood the region as the county in which they lived, whereas 14 percent saw it as the geographical region (ZMP 2003). In the German context, the term regional is also closely linked to idea of "cradle to grave," where the life cycle of the product is contained within a specific area (Kindermann 1997). Thus a clear definition or common cultural understanding of what is local can never be achieved because it is spatially and culturally embedded in the context of its use.

Consumers of local food can be defined by a shared wish to strengthen a connection to a commons, to land and landscape, to people and place. According to actor-network theory, strategies that support relocalization are based on building production-consumption networks around specific meanings and objectives. The success of such strategies depends on its capacity to align actors along shared axes of meaning as the condition to coordinate their action (Latour 1987). In terms of the context of the market, farmers' markets work at achieving such a reconfiguration, since they seek to create an "immediate, personal and enacted in shared space" (Lyson and Green 1999, 137). Such direct market venues, which alongside farmers' markets also include community-supported agriculture and vegetable box schemes, create links between the farmer or producer and the consumer through simple everyday exchange. According to Clare Hinrichs (2000, 295), "direct agricultural markets promise human connection at the place where production and consumption of food converge." Embeddedness, in this sense of social connection, reciprocity, and trust, is often seen as the hallmark of direct agricultural markets. The social and spatial grounding of food in a local system enables and constrains production and consumption through its own unique characteristics (Hendrickson and Heffernan 2002). In direct counterpoint to a globalized food marketplace, embedded markets leave the supply of food open to local agency and change. It is at this scale that differentiated and discernible differences in the spatial and temporal experience are fundamental to the concept of local food.

In this chapter we will look at the emergence of an interest in re-localizing food systems. The "re-placing" of food can be seen not just as reconstructing relationships between consumer and producer, but as a way of people re-embedding their relationship with food within a sense of the spatial structure of their lives. We investigate this through two stages: first, we review of the concept of localism and food and how this constructs different relationships between people and place; second, we put this concept into context through a qualitative case study of the use of a German-based

Web resource, www.mundraub.org. The mundraub.org website invites people to tag forgotten fruit trees on an interactive map and to locate existing trees that can be harvested. We will explore how the crowdsourced creation of a map-based resource documenting over five thousand fruit trees and wild edible plants can create a redefined sense of the local and re-embed food consumption within a local spatial setting. Interestingly, mundraub.org is widely used by urban dwellers and a large proportion of the map tags relate to urban and extra-urban tree sites, so this re-embedding should be seen within the context of a re-examination of the urban and its relationship with food.

Scrumping, Foraging, Scrounging, and Collecting

Farmers' markets, local co-ops, and vegetable box schemes represent the direct link between producer and consumer. It is for this reason that they can be seen as a fundamental basis for the local food movement. At the same time, a reemergence of other forms of local food sourcing is taking place, specifically foraging and scrumping. In these activities there is no person at the production end of the supply chain; instead, there is only the consumer, in direct relationship with the source of the food. Thus the consumer is also the farmer, in the sense of harvesting the food and preparing it for eating. In this way the natural world offers a significant counter to the idea of a market based on monetary exchange: wild flowers, trees, and herbs offer a free exchange, where ownership is ambiguous and rarely contested. The scrumping movement takes a wider approach to forming a relationship with the natural world and with the idea of "wild" food, one in which it is important to understand how the relationship with nature and the sourcing of food has shifted.

The website platform mundraub.org supports scrumping, and is primarily active in Germany (Frosch 2012). The German term "Mundraub" is grounded in that country's tradition of planting trees as a resource. For instance, up until the nineteenth century people were obliged by law to plant a fruit tree when they married (Krug 1808). Initially there was no law preventing commoners from helping themselves to the fruit, but in the build-up to World War II a law known as "Mundraub"[1] was introduced, which roughly translates as "stealing from the mouth" (the closest translation of the practice is the English term "scrumping."[2]). The law made taking produce from these trees illegal. After the war, more trees were planted in East Germany due to a widespread fear of food shortages, and the Mundraub law was only withdrawn in 1975. But even now the harvesting of fruit in Germany is illegal unless the owner gives permission, and failure to get consent can potentially leave the individual liable to being prosecuted for theft.

This highlights an important and charged aspect of scrumping as an activity. There are effectively no fruit trees in Germany that are not owned: trees either belong to private persons or to the state, county, or local parish, depending on who owns the land where they grow. For this reason, scrumping is often regarded as a semi-illegal activity and equated with poaching. The practice is also culturally situated, with different connotations in different cultural contexts. In Germany and in the United Kingdom it is tolerated and a widely known practice, but scrumping is fairly unknown in the United States, where the nearest equivalent term is "rustling." In Sweden it is referred to as "palla," specifically meaning to steal fruit from trees in other people's gardens. The practice of scrumping has particular resonance in German society, however, which has a tradition of seeing the natural landscape as something that is not just passively there, but has the capacity for "giving." It is understood by many as restorative or healing, and recognized as a resource. For example, the German field titled "Naturheilkunde" (Naturotherapy) sees nature providing the opportunity and capacity to set free "bodily energy and spiritual forces which can instigate self-healing" (Brauchle 1935, 153), where one of the key practices is the consumption of raw food. Complementing this is the idea that the countryside and nature can offer healing through plants and natural medications, so that the rural world is traditionally seen as the "farmer's apothecary." Yet the current wider global progress toward supermarket food and shopping has meant that the necessary local knowledge about the qualities of particular plants or food is no longer a part of everyday life, particularly for the city dweller but also for rural residents. The revived interest in scrumping can be seen in the context of a desire to find ways of sourcing local food directly from the natural world.

The economist Avner Offer (1997) proposes a framework for an understanding of nonmarket exchange that introduces the idea of the "economy of regard." These transactions are usually referred to as "gifting" and take place in the context of non-monetary reciprocity. The gift can be expensive or cheap, substantive or symbolic, but it is not costless. Here the value of the transaction is not measured quantitatively but in terms of "intrinsic benefits of social and personal interaction; from the satisfaction of regard" (Offer 1997, 450). Regard can take many forms, but at the very basic level "regard" is a grant of attention, underplayed by the motivation of the mutual benefit of sharing goods. This is important notion for the interpretation of the practices of scrumping, foraging, and scrounging, since regard in this context is not exchanged between two people but instead directly between some natural source of food and the potential consumer. The regard is then not a social regard between two humans but a direct reciprocity between the resource (which is seen as gifting or giving some fruit or herb) and the person (who harvests the fruit or herb for their consumption).

We now discuss a qualitative study of the use of the mundraub.org platform, and try to understand how concepts of localism and the corresponding relations of regard are constructed and practiced.

Case Study of Mundraub.org

Mundraub.org creates a platform through which wild or forgotten fruit trees, as well as nut trees, herbs, or berries can be identified by one person or group, and then located and harvested by others, with an open-access interactive map. The mundraub.org platform is a simple map with tags that can be created by any user to add a food source categorized as fruits, nuts, berries, herbs, and gardens. Justin Buckley, Katharina Frosch (co-author of this chapter), Kai Gildhorn, Daniel Nielsen, and Mirco Meyer initiated the platform in 2009 following a canoe trip to the former East Germany, during which they noticed large numbers of forgotten fruit trees. It is a nonprofit site currently run by volunteers, and to date there have been over seven thousand five hundred tags created (status as of July 31, 2013), many of which denote multiple trees, orchards, and plants so that the individual tags represent a much larger number of trees and plants. A tag will list an entry according to a category and subcategory (e.g., fruit, apple) and the entry format prompts for a "who, where, what, how" description of the find that often includes a text description with directions to the location—see figure 9.1 for a sample tag. The site does not provide details such as GPS coordinates or use other location-based technology, and it is not currently available on a mobile device (though a mobile platform is planned for 2014).

Typical content is focused on the description of the fruit and a description of how to find the tree or bush. This is often augmented with other details, such as the taste of the fruit. There is an associated Facebook site, which has 4,762 members (status as of July 31, 2013). Typical entries peak in autumn and focus on identifying fruit types, sharing of recipes for preparing or preserving fruit, and general information about related events or projects nationally—see figure 9.2.

Method

We studied the use of mundraub.org using a range of methods. Initially we undertook a data-mining study of the mundraub.org website entries and also the parallel Facebook group site. The main qualitative analysis was through a series of semistructured interviews of approximately thirty minutes with eight mundraub users. We recruited eight mundraubers from Germany, mainly through an advertisement on the Facebook website. There were six men (Ola, Michael, Sebastian, Marco, Niko, and Karl) and two women (Simone and Danielle) with an age range from the twenties to sixties. None were paid for their participation and the interviews were undertaken in German (thus, any quotations in the following text are the authors' translations of the original response). The participants came from a range of urban and rural home locations and covered a range of mundraub.org experience. In selecting participants the aim was not to create the basis for statistical comparisons across different types of users and

Figure 9.1
Mundraub tag in Berlin. Tag: "Far too many apples. Right on the road." Cibola comment: "Loads of apples, ideal for juice or apple sauce."

experience levels (which would not be appropriate with such a sample size). Rather, the aim was to provide an opportunity for issues particular to different types of users to be raised in the study.

Findings

In this section we present a subset of findings from the qualitative study. We organize our findings around the three different ways in which people use mundraub.org: first, the descriptions from the interviews of the individual experience of finding, tagging, and harvesting fruit; second, the characteristics of the tag; third, the way in which the Facebook site was used to share experience and document scrumping outcomes. We consider the findings in the context of how ideas of "locality" were practiced and how relations of regard were expressed and acted out.

The main common experience expressed by those interviewed was the way in which people started to develop a different awareness, understanding, and appreciation of what was local to them. This was almost always equated with "seeing" the place differently, but also related to the seasons and the change in their experience of time. The common aspect of the way the participants related to the fruit they had harvested was an interest in finding out about how to prepare and preserve it, which they found they needed to learn.

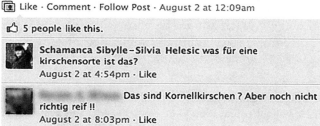

Figure 9.2
Facebook discussion around the identification of the fruit pictured in the image. (Schamanca, "What sort of cherry is this?" Answer, "They are cornel cherries? But not ripe yet!!")

Awareness

Interviewees commented on how the use of the mundraub.org platform had opened their eyes to an awareness and identity with their local environment. This was independent of whether they were city or rural dwellers. They saw this in contrast to other inhabitants who "don't seem to know what there is in their local environment" (Ola). They also talked about the way in which both everyday and special journeys were somehow woven into the mundraub.org exploration. For example, one commented how "when I go walking I just look around" (Michael). Others described how a car trip was interrupted by the discovery of fruit trees along the road: "We were driving and we just saw all these trees so we stopped and gathered damsons" (Ola). Another described how his frequent train trips became opportunities to spot apple trees along the train tracks. The heightened awareness and practice of looking is underlined by one participant who described how:

I always try to move through the city with open eyes, and am always pleased when I find a tree with fruit somewhere. . . . This definitely changes my perception of the city. I associate certain places with certain trees and seasons. I also discover new areas of the city that otherwise I might never have been to. (Sebastian)

All of the interviewees referred to the practice of watching trees in their locality, and this being not just a one-off relationship of observation but a longer-term involvement with the tree location and a consequent sense of themselves being intertwined with the fruit tree places.

Seasonality

The practice of watching for what was happening both around them and in certain locations was finely linked with the seasons and thus a subtler connection with the temporal changes in a place. An interviewee who had many years of experience with scrumping and who lived in an area abundant with abandoned trees talked about becoming so finely aware of the natural processes that he "watched for when the bees start to fly" (Karl), since the first step toward a fruit tree bearing fruit is the pollination by bees. Another noted how "you get into the mindset of thinking when the fruit will be ripe" (Ola). Another commented on how the experience made him curious:

It's really interesting to watch the fruit trees and wonder is it ripe or unripe. I start to look if the fruit is getting ripe and start thinking about when it might be ready to pick. This year I really looked forward to apple picking time. You find yourself following the trees' development. (Michael)

All of the interviewees noted how the dormant period of winter left a gap, with one commenting how "in winter nothing happens, well it was like that last winter anyway. It's a shame" (Marco). As well as experiencing an appreciation and awareness of the seasonal affect on the fruit trees in terms of the fruit tree's development, interviewees also thought about the life of trees over the long term. They developed an awareness of the history of the tree and an acceptance of a cyclical process repeated every year. For instance, one interviewee noted how "when we come across a tree where fruit was lying on the ground and just rotting then we make sure that we go there a bit earlier the next year" (Karl).

This aspect of timing is often highlighted in the mundraub.org entries, for instance with a note in the tag about when the fruit was ripe: "Harvestable from the end of July (try first). In season until end of September, from October onwards it is a bit of a risk since there is hardly any fruit left on the tree" (Sunday, July 25, 2010, Dalem). Similar to the relationship with place is the relationship with the seasons, and thus with time. This suggests both a heightened awareness of the embeddedness of the tree's development (and thus usefulness) in time and also on a wider level—in the history or environmental changes in the place and the connection between this change and the tree's placement in the landscape.

Appreciation and Regard

All of the interviewees expressed the perception that it was just a shame not to collect fallen fruit and make use of it. For example, several people echoed this prevalent feeling by saying, "I just think it's a shame when good fruit just drops off the tree and it isn't used" (Ola), and "there is so much delicious fruit that grows near us and that no one uses" (Marco), and "for so long it has just been there for the birds" (Michael). This experience was described in the same way by city dwellers and town residents as by those who lived in the countryside. One interviewee described his experience with a pear tree in his own garden: "We went away for a week and by the time we got back the fruit was no good—I was a bit annoyed" (Michael). In fact, the frustration at seeing "unused" trees was often most strongly described by those who lived or had links to the country, where there were many "abandoned" trees or orchards in which they perceived the fruit was going to waste. One interviewee (Michael) described his visit to his parents' village where "there are lots of fruit trees which are hardly ever picked, and then I noticed in my parents' fridge that the apples come from the supermarket." This reporting of "unused" fruit (i.e., fruit "on offer" but not being taken) shows how the relation of regard between the tree and the person suggests that not harvesting the fruit would be in some way disrespectful. The idiom that an older interviewee who lived in a small town close to the country where lots of fruit trees were abandoned recounted was that "what grows has worth" (Karl). Karl talked about how he and his wife "were given a field pretty much as a gift by the owner, not the field itself, but the fruit in it." The trees, through their gift of fruit, can be seen as creating an obligation to reciprocate (Offer 1997, 455). All the participants expressed some obligation to harvest the fruit, and that by doing so they felt rewarded in a manner totally outside of monetary value. In fact, people went out of their way to emphasize that they didn't scrump for financial reasons. Yet scrumping did not, in most cases, replace or substitute the mundraub.org users' consumption of fruit from shops. They still reported buying fruit during winter, but with a different awareness of the fruit as product; some recounted that they subsequently only bought fruit grown in Germany, or choose certain older varieties of apple. Thus scrumping was not viewed as an economic activity, but was driven by a compulsion to reciprocate and accept the gift that the tree was offering, where the value was still embedded within the context of the particular tree or orchard where the fruit was harvested.

Sharing

The need to harvest "unused" or "forgotten" trees sometimes meant that people gathered much more than they could practically consume, and therefore they had to consider how to store, preserve, and prepare the fruit. In this sense the harvesting of the fruit became a point of negotiation with a wider social group. It opened up the process of gathering the fruit to a possibility of sharing what had been harvested. This happened on many levels: first, there was the sharing of knowledge and experience; second, the literal sharing or offering of surplus fruit; third, the sharing of recipes and modes of preserving or preparing fruit; and finally, the possibility of literally sharing the act of harvesting the fruit. Since the mundraub.org site is anonymous and provides only map-based tagged information, the Facebook site and links within a person's own social community became the framework through which this sharing or exchanging practice was enacted. Sharing place-based information can be a rich social practice and serve as the basis for creating a sense of communal co-presence (Willis et al. 2009). Interestingly though, there is no social network structured around mundraub.org usage. Although there is a large number of members on the Facebook group there is little differentiation between members and no explicit links between friends. When people tried to establish these social links, they reported how they been unsuccessful. For instance, one interviewee, who lived in a large city, described how he put a message on Facebook asking if anyone wanted to come and pick damsons because "there was a space in the car and it seemed a shame not to share. But no one replied. I don't know why" (Ola).

On the Facebook site one interviewee posted the message "who will take some of my quinces?" (Danielle), which resulted in two replies, one being "if you had plums I'd happily help" and the other being "Frankfurt is just too far for me." None of the interviewees reported using the Facebook site in conjunction with their existing social networks as an alternative marketplace. Thus their motivation to share was thwarted by the geographical distribution of the Mundraub members on the Facebook site and a lack of interest or engagement with an alternative place of exchange in their own local social network.

This suggests a very different approach to foraging and hunting as a practice, since although there was a motivation to share individual finds with a wider group there was actually no demand for this. In terms of understanding how scrumping through mundraub.org is constructed as an non-market exchange, it is important to note that it does not operate through the commonly accepted model of the benefit of communal sharing or through the joint exploitation of a resource, as is often practiced in foraging (Offer 1997, 457). Neither does it specifically support a regional or localized approach to the fruit: there are no regional subgroups of members and knowledge about fruit identification, and preparation is not region-specific. In this sense the Facebook site, although constructing some form of social structure for sharing local knowledge, does not actually allow for this knowledge to be re-embedded locally since the site itself is globally accessible.

Harvesting

The main encounters that participants had at a social level occurred at the point when they picked the fruit. Here many reported funny stories about the public perception when they started to harvest fruit trees. This was closely tied in with the negotiation of the "ownership" of the tree. The process of observing a suitable tree and then trying to find out who owned it—either an individual or a local body—and if it was a "forgotten" tree became one in which all Mundraubers necessarily became involved. Many chose to literally avoid people and human contact when harvesting a tree to avoid confronting any issues about ownership. For example: "I used to just see a tree and when I had time and there was nobody about or thought that the tree didn't belong to anybody I just picked the fruit. But now I am much more prepared—I find out first and then pick" (Marco). The practical aspect of trying to ascertain the ownership of the tree is documented in many of the mundraub.org website tags. A large proportion of the tags make some reference to ownership. Some describe this based on local knowledge of the trees or orchards and some based on observation of whether the tree was within some form of boundary such as a fence. Yet one interviewee highlighted an awareness of the fine line that scrumping walks between legal and illegal activity when he stated, "We are a little but like thieves" (Karl). Others described how created an audience when they did pick fruit in a place with passersby. One city dweller described an unintentional "public performance":

A highlight this year was to discover a damson tree at the U-Bahn [underground or metro] station just round the corner from where I live. When I picked the fruit I used an apple picker, which you really need in the city, and I got quite an audience with people watching me who were really quite surprised about what I was doing. I ended up answering lots of questions mainly from old ladies, but generally most people were very shy and reserved about the whole thing. (Michael)

Another described a confrontation he had while picking a tree in the city:

People's reactions when they see you picking a tree can be really varied. Most seem quite interested and ask what you are doing and whether you can eat the fruit. But I had another situation last summer. I harvested a branch with Amelanchier [shadbush] berries in a residential street. It was next to a furniture shop and the owner came out and was really interested. I let her try a few berries and she then went independently and looked on the Internet to see what sort of berries they were. From the same building an old lady appeared and complained bitterly that I had picked "the beautiful berries" from the tree, and that it looked ugly without them, and that I should seriously consider if that's what I should be doing. I offered her the berries to try but all she was interested in was what they looked like. (Sebastian)

Despite the different contexts for the trees—the single tree in the urban setting, the extra-urban orchard or the line of trees along a country road—the negotiation of ownership at the point of harvesting and the confrontation between people with different understandings of the use of trees were common experiences to all interviewed. This demonstrates that the perception is not differentiated between the urban or rural context, but more by a sense of people paying attention and valuing the resource within the locality. For instance, a participant described how "one day, we started harvesting some cherry trees. And some of the people living or working nearby were really stunned because they had never realized that there were real cherry trees. One lady said: I have been working here for thirty years, and I've never noticed this cherry tree that grows directly in front of the shop door" (Kai). The "noticing" or paying attention to the tree was something that was undertaken by the Mundraubers. Other locals were either simply not aware of the tree or had a different perception of it, such as a visual appreciation of the berries, but not as a source of fruit for consumption. At the point when the Mundrauber "paid attention" to the tree while harvesting it, the tree was transformed in terms of how it was perceived. It became a site of discussion; a meeting of these different perspectives, and sometimes a confrontation where different value sets were applied.

Defining Local

The final aspect of our findings was the way in which a tree was defined as being local to the interviewees. This occurred on two levels; first they described a certain distance from their home that was local. All the interviewees described this as being "about five kilometers." A rough check of participants' tags in relation to the area in which they live confirms this.

The locality of the tree was not literally on their doorstep but it was located within a circle of about a five kilometer radius, which extended to, and tended to coincide with, the land that lay on the edge or just beyond the perimeter of the village, town, or city in which they lived—see figure 9.3. "Local" was also defined through the quality of the fruit. In the study of the tags for the website a large number gave some description of the visual appearance and taste of the fruit. For instance: "The pear is small and very juicy, crunchy and sweet," or "tangy-sour, but still good to eat without having to pull a face :-)." Another interviewee described how he had taken the initiative to bring the fruit into his children's kindergarten "just to let them know about what fruit was growing nearby and let them see 'real' fruit" (Michael). Another pointed out how wild fruit is more distinctive in several respects:

Apples from wild trees just look different. Gala apples always taste the same. It doesn't matter when or where you buy them. They look the same too. But on wild trees the apples tend to all look different; they're not perfect for a start. Also wild fruit trees tend to be a whole range of apple varieties." (Marco)

Thus, "local" in this sense was a much more differentiated experience of the fruit. Each tree and location had specific characteristics, which were only definable through the direct experience of tasting the fruit; most would find a tree and pick the fruit and try it for taste. In fact, despite their interest in their local trees and the quality of the fruit, only one interviewee felt able to define himself as an expert on apple varieties or identifying fruit. Indeed, one interviewee who had also scrumped for many years,

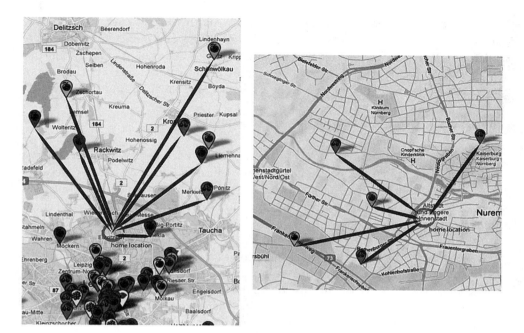

Figure 9.3
Fruit tree tags and the distance relationship to two Mundraubers.

even before using mundraub.org, claimed to know only three or four fruit varieties by name and sight. Instead, he knew and remembered individual trees and orchards and what the fruit tasted like and its particular usefulness: "One type you have to eat pretty much straight after picking, another we can keep until February" (Karl).

Interestingly, a number of interviewees set the context of their scrumping activities not just in terms of where they currently lived but also in relation to the place where they had grown up. Some had made entries in both locations, and one commented: "I also found trees in the village where I grew up," whereas another described how she had "grown up in the country and my parents and grandparents had fruit trees."

A final way in which locality was defined was through the description of the tree in the tags on the website. Rather than simply describing the physical location and how to find the tree, the people often posted some context for how the tree came to be there in the first place. Some conducted a "fruit tour" by tracing the route to the tree through other places, for example, "an old apple avenue in the middle of some fields . . . whatever you don't like just give to the horses in the opposite field; nearby on the 'Apothecary-hill' there is a viewing tower with a great view of the surroundings." Others describe the history of the place and why the fruit trees are now abandoned such as many "fruit tree meadows."[3] Another entry noted, "On the way [are] lots of apple and pear trees as well as plum, cherry, and quince trees. These are what are left over from a village that became victim to the mining industry." In this way locality was not just defined as a geographic territory or boundary but as a set of tastes and appearances of certain fruit or a sense of the spatial locality as history.

Discussion: Embedded Relationships with Food and Place

The key characteristic of mundraub.org use was that it documented an embedding of a relationship between people and place. If we return to Offer's concept of "regard," this can suggest an approach to understanding the nature of this embedded relationship. Offer (1997) refers to the grant of attention, motivated by the mutual benefit conferred by the process of sharing. This can also be extended to the way in which all of the interviewees in the study described their experience as being about gaining a new "regard" for the place through the attention paid to it in the scrumping activity, and particularly an awareness of the locality and seasonality. Offer explains this type of exchange as a "process benefit," usually in the form of a personal relationship (1997, 251). This was most clearly highlighted in the shared motivation of not wanting to waste good fruit, but also more indirectly in the changed perception of time and seasons at a local level. Through the attention paid to the trees the Mundraubers became sensitized to the temporality of local fruit; there was a palpable expectancy that spring would bring a new year of fruit and that certain trees were ripe at certain times. This was recognized as valuable by the interviewees and was seen as something gifted to them in the process of exchange between tree, locale, and the harvesting of fruit. There was a heightened sensitivity also reflected in the way that individual trees were understood as having fruit that had distinctive characteristics and properties. This direct experience of the tree was literally negotiated through taste. In this way such food products become embedded through a process of mobilization of values and meanings that construct a place as the "local." According to Roberta Sonnino (2007, n.p.) this localization process is about creating a discourse that "roots this product in time and space." This concept of embeddedness is highlighted by Ash Amin, who states that, "the local emerges primarily as a 'relational space,' a space of interrelated scales and interdependent subjects where the social, economic, political and cultural inside and outside are constituted through the topologies of actor networks" (Amin 2004, 33). The link, acknowledged by both interviewees and in some tags, between the tree and the history of the place or the birthplace, reinforces the embedded nature of the relationship with the tree, not just in the current time, but also in the past history of the place.

Mundraub.org also created a new granting of attention and exchange to forgotten, in-between, and non-places as valuable sites and possible places of consumption and exchange. This involved recognizing a whole range of sites as sources of fruit: trees along the side of roads, in car parks, and in the middle of urban sites not typically viewed as "natural." This also included a number of links to abandoned and underused places, which through the location of productive fruit trees were not perceived as derelict or "broken" but as sites of growth and difference (Franck and Stevens 2006). The discovery of fruit trees in city-center locations was an important aspect to city dwellers; a Berlin resident commented, "Before Mundraub I didn't realize that there were openly accessible fruit trees right in the middle of the city. I appreciate Berlin all the more now because of this. I look much more carefully at the trees in parks and in the streets" (Simone). Other interviewees described how they started to reassess the functionality and value of locations previously regarded as non-places, such as transport hubs and car parks. For example, one noted, "Sometimes they're in places you'd never expect—like car parks. You make use of the place so much more" (Michael). Thus the attention to the local in this sense was a place around which a person could construct meaning, since it was not regarded as being owned or within an existing community's territory. In this way local eating can be understood as being embedded in small-scale, highly differentiated spaces that have specific qualities and territories. This is in stark contrast to the homogenized global spaces of fast food takeaways and chain supermarkets. By reinstating small and disregarded sites as locations of value, opportunities for a new sense of relatedness in urban space were opened up.

Summary

We discussed how changing ideas of locality and place in relationship to food affect an understanding of the production and consumption of food. In this context, the "re-placing" of food can be seen not just as reconstructing relationships between consumer and producer, but as a way of people re-embedding their relationship with food through a sense of locality. In our case study of a scrumping Web platform in Germany, we explored how this locality is constructed through four perceptions: a simple awareness of trees in the environment; a changing sense of time and seasons; the exchange created through picking fruit; and a refined sense of what local meant in spatial terms. These changing relationships with place were enacted through a "regard" for, or by paying of attention to, the tree and the place, and with a reciprocal exchange that created a sense of embeddedness.

Acknowledgments

We would like to thank the interviewees that participated in our study: Michael, Sebastian, Marco, Ola, Simone, Karl, Danielle, and Niko. Thank you also to Daniel Nielsen and Kai Gildhorn from mundraub.org.

CINNAMON PLUMS

The stewed plums can be served with pancakes, vanilla ice cream, or rice pudding. If stored in a cool, dark place or in a refrigerator it will keep for a few months.

Ingredients

2¼ lbs. ripe plums or damsons

1 cup water (or alternatively, red wine)

1¾ oz. caster sugar (known as "baker's sugar" in the US, it is more finely granulated than table sugar but not as fine as powdered sugar)

2–4 cinnamon sticks (ideally large cinnamon sticks, which are slightly more bitter and tangy than the smaller ones)

Preparation

1. Wash the plums or damsons, halve them, and remove the stones. Place the fruit in a large wide, heavy-based saucepan with the water, sugar, and cinnamon sticks. Cover and bring to a boil.
2. Cook gently over a low heat, trying to avoid stirring the fruit. Once the fruit is tender but not falling apart, remove the pan from the heat.
3. While the fruit is cooking, sterilize preserving jars and lids by boiling them in water. Ladle the hot stewed fruit into the hot jars, cover with the lids, and carefully turn them upside down. Wipe any spillages with a clean damp cloth and leave the jars to cool completely.

Notes

1. See http://de.wikipedia.org/wiki/Mundraub.

2. Scrumping has a slightly different cultural connotation. The practice of "Mundraub" is legally considered theft if the owner is not asked for permission beforehand, but it is seen generally as an opportunistic and publicly acceptable activity.

3. The German term is Streuobstwiese, which refers to a traditional type of grassland-orchard management system consisting of a meadow with scattered fruit trees. They have now become rare; over 70 percent of meadows were lost between 1965 and 2000.

References

Allen, Patricia. 2004. *Together at the table: Sustainability and sustenance in the American agri-food system*. Philadephia: Pennsylvania State University Press.

Amin, Ash. 2004. Regions unbound: towards a new politics of place. *Geografiska Annaler* 86B:33–44.

Brauchle, Alfred. 1935. Was ist der naturheilkunde? *Der Naturarzt* 63:153–156.

Duram, Leslie, and Lydia Oberholtzer. 2010. A geographic approach to examining place and natural resource use in local food systems. *Renewable Agriculture and Food Systems* 25 (2):99–108.

Franck, Karen, and Quentin Stevens. 2006. *Loose space*. London: Routledge.

Friedmann, Harriet. 1993. After Midas's feast: Alternative food regimes for the future. In *Food for the future: Conditions and contradictions of sustainability*, ed. Patricia Allen, 213–233. New York: Wiley.

Frosch, Katharina. 2012. Mundraub? allmendeobst! In *für eine neue politik jenseits von markt und staat*, ed. Silke Helfrich, 273–275. Bielefeld: Transcript Verlag.

Hendrickson, Mary, and William Heffernan. 2002. Opening spaces through relocalization: Locating potential resistance in the weaknesses of the global food system. *Sociologia Ruralis* 42:347–369.

Hinrichs, Clare. 2000. Embeddedness and local food systems: Notes on two types of direct agricultural market. *Journal of Rural Studies* 16:295–303.

Ilbery, Brian, and Moya Kneafsey. 2000. Registering regional speciality food and drink products in the UK: The case of PDO's and PGI's. *Area* 32 (3):317–325.

Kindermann, Arndt. 1997. *Okologische chancen und perspektiven von regionalproduktion und regionalvermarktung. Naturschutzbund Deutschland e.V*. Bonn: NABU.

Krug, Leopold. 1808. Geschichte der staatswirthschaftlichen Gesetzgebung im preußischen Staate: von den ältesten Zeiten bis zu dem Ausbruch des Kriegs im Jahre 1806, Band 1, Berlin.

Latour, Bruno. 1987. *Science in action. How to follow scientists and engineers through society*. Cambridge: Harvard University Press.

Lyson, Thomas, and Judy Green. 1999. The agricultural marketscape: A framework for sustaining agriculture and communities in the northeast US. *Journal of Sustainable Agriculture* 15 (2/3): 133–150.

Offer, Avner. 1997. Between the gift and the market: The economy of regard. *Economic History Review* 50 (3):450–476.

Oltersdorf, Ulrich, and Kurt Gedrich, eds. 2001. *Ernährungsziele unserer Gesellschaft: Die Beiträge der Ernährungsverhaltens-Wissenschaft. Berichte der Bundesforschungsanstalt für Ernährung, Karlsruhe. Dokumentation zur 22.* Bonn: AGEV-Jahrestagung.

Paxton, Angela. 1994. *The food miles report: The dangers of long-distance food transport.* London: SAFE Alliance. http://www.sustainweb.org/publications/?Id=191.

Sonnino, Roberta. 2007. The power of place: Embeddedness and local food systems in Italy and the UK. *The Anthropology of Food* S2 (March). Available at: http://aof.revues.org/index454.html.

Willis, Katharine, Kenton O'Hara, Thierry Giles, and Mike Marienek. 2009. Sharing knowledge about places as community building. In *Shared encounters*, ed. Katharine Willis, Konstantinos Chorianopoulos, Mirjam Struppek, and George Roussos, 291–308. New York: Springer.

ZMP. 2003. *Nahrungsmittel aus der Region—Regionale Spezialitäten.* Bonn: ZMP.

GROW

Eli Blevis

The six chapters that make up this section on growing are quite diverse in their perspectives, backgrounds, and styles and yet unified by a concern for food-related issues of scale and locality, authenticity and verification, sustainability practices and futures. Each of these chapters pushes the boundaries of what we think of as human-computer interaction (HCI). In connecting HCI and growing as a special concern of food, the authors collectively illustrate that design inquiry in HCI is particularly effective when it is taken to be a domain-centered concern.

In this age of climate change, food represents one of the most dynamic concerns facing humankind. Where we can grow food and what we can grow will change in the decades that are ahead in an unprecedented way. Figure III.1 accompanying this introduction is my own image of a wind-powered growing installation in the Central Business District of Hong Kong. The small box at the left of the installation shows passersby how much energy the wind turbines create, and the green patch of growth stands in sharp contrast to the urban surroundings. This installation emphasizes the need for thinking differently about how and where things grow. Yet there is something disturbing—intended or not—about the choice of a patch of grass as a mechanism of social awareness about energy use. As pictured, the installation represents a failure to connect the need to grow food, rather than lawns, to the need to be more sustainable in our energy production. The chapters in this section have many nuances and collectively denote a different way of seeing and understanding vital importance of growing to our future. They also include many tangible design concepts that look ahead to a secure, safe, and ethical food supply, and thus prescribe a direction for HCI crucial to our collective future.

In "'You Don't Have to Be a Gardener to Do Urban Agriculture': Understanding Opportunities for Designing Interactive Technologies to Support Urban Food Production," Will Odom covers the topic in three "narratives": (1) an account of motivations within sustainable and HCI literature for a focus on food; (2) a report on fieldwork, including shadow studies, and interviews conducted over eight months in Australia with the urban agriculturalists from two community garden sites; and (3) a description

Figure III.1
Wind-powered installation in central Hong Kong

of several ideas for supporting small-scale food production with interactive technologies. Odom's findings include an articulation of the tensions between local community food production and local government policies. By way of design implications, Odom suggests—after Tony Fry—that food waste may be "re-coded" as a strategic resource (i.e., as compost which promotes more food growth). Odom also describes possibilities for constructing interactive systems to promote resource exchange among growers, and interactive systems to make growing more visible.

In "Augmented Agriculture, Algorithms, Aerospace, and Alimentary Architectures," Jordan Geiger considers the contemporary phenomena of "very large organizations" (VLOs), including "big" agriculture, and the notion of the commons and failures of the commons. Geiger also describes a project with Virginia San Fratello called Hyperculture—Earth as Interface, in which the architectural gaze is focused on the "production of landscape rather than building," and which pushes beyond monoculture notions of growing and beyond polyculture to propose a Google-Earth styled, digital-world, sensor-enabled hyperculture. Hyperculture includes a site called CSA (community supported agriculture), which enables farmers and consumers to participate one with another directly, sharing risks and deciding what to grow.

In "The Allure of Provenance: Tracing Food through User-Generated Production Information," Ann Light takes up the issue of food tracking as a matter of sustainable and ethical practices in the guise of a project she calls The Fair Tracing Project. Light recounts her experience with coffee growers in Karnataka, India, to illustrate her notion of food *provenance*. She articulates provenance's potential as a way of marketing ethically produced food. She also takes up the issue of establishing trust in the description of provenance—an important counterpoint to this potential. Figure III.2 accom-

Figure III.2
Farmer's market at the crossroads of a Hong Kong pedestrian walkway

panying this introduction is my own image of a tiny farmer's market on a walking platform in Hong Kong in which placards at the front of tables describe the provenance of what is sold. Light tells us: "The smallest producers, delivering tiny batches of coffee, cannot maintain the identity of their beans, however good the quality. Their crop—and, with it, detailed provenance—is lost into a pile of similar beans." In response, Light's notion of *provenance* is one of political and ethical imperative. She explains the project's origin: "The Fair Tracing tool was conceived as a means of leveling the playing field between small producers and big companies who are able to commission tracking technologies and exploit product data (usually without sharing it)." With great honesty, reflection, and candor, Light shares with us her evolved understanding of her story of the Karnataka coffee growers by writing, "The description at the start of this chapter, then, is my story, not that of the growers; it is the record of my distress." Light closes with a more optimistic account of how economies of exchange may help create more equality and value than economies of production and consumption.

In "Beyond Gardening: A New Approach to HCI and Urban Agriculture," Tad Hirsch focuses on the opportunities for HCI research to target urban agriculture and the importance of doing so as a matter of developing sustainable practices. Hirsch offers a definition, a framework, and a case study to advance this focus, asserting that, "Nowhere is the blurred boundary between rural and urban space more apparent than in food production." Figure III.3, my image accompanying this introduction (as well as with a detail from the photo in figure III.4), shows factories in southern China's Pearl Delta region encroaching on and blurring the landscape between agricultural and urban. Hirsch asks us to think beyond cultivation to a more systemic notion of urban food provisioning. This reframing leads to Hirsch's definition of *urban agriculture* as "the practice of producing and distributing food within cities, in ways that reflect the

Figure III.3
Farming aside factories in southern China

Figure III.4
Detail from figure III.3

distinctive character of urban life." Four key aspects in urban architecture also emerge from and frame Hirsch's analysis, namely *cultivation, distribution, organizations,* and *culture.* Hirsch applies this framework to the case of Sunroot Gardens—a CSA project in which he participated. The framing helps him to conclude that although Sunroot Gardens "engaged in cultivation, its aim was to reinvent entire systems of food provision." From the technology perspective, Hirsch asserts that the opportunities for HCI include the design of "technology artifacts to reduce dependencies on fossil fuels, build resilient networks, simplify supply chains, mitigate the influence of money, and promote biodiversity."

In "Hungry for Data: Metabolic Interaction from Farm to Fork to Phenotype," Marc Tuters and Denisa Kera take up a notion they call *metabolic interaction,* which they define (paradoxically and poetically) as "to eat and to know what we eat, where it comes from and how it interacts with our molecules, to be at once a beast and a god." Much of Tuters and Kera's contribution takes up—as Hirsch does with rural and urban—the difficulty of discerning local from global; as they explain it, "the local and the global are both essentially fictions." Tuters and Kera give a wide-ranging discourse about all things related to food——and how food might be constructed within HCI and beyond—with less emphasis than some chapters in this section on growing but with no less of a compelling message. The chapter is so rich in its referencing that it may be partaken as a kind of highly informed and nuanced dinner conversation. At one point, for example, Tuters and Kera write: "Eating is as much a social, cultural, and political act as it is a natural act, happening on a table as a place where discourse, nature, and culture meet to connect our animality with what defines our civilization." Tuters and Kera articulate particularly clearly an assertion that concurs profoundly with those made by authors in other chapters: "In our design research we take seriously the ontological implications of new technologies and scientific 'discoveries' in terms of how they might impact conceptions of human subjectivity and community." To this end, Tuters and Kera offer technologically material scenarios that are ontologically articulated between notions of being and becoming, with, namely, GoodGuide.com and the DIY SousVide cooking appliance on the one hand, and the social media site FridgeMatch and the prototype for a genetically directed meal in "Eat What You Are" on the other.

In "Food Futures: Three Provocations to Challenge HCI Interventions," Greg Hearn and David Lindsay Wright describe a taxonomy they call the Food Futures Cube as an instrumental catalyst to HCI design related to food. To set the stage for the Food Futures Cube, Hearn and Wright provide a rich account of visions of the future that are "technotopian" in contrast to visions of the future that are not focused on the continuous creation of more and more, but rather promote a return to a simpler life, focusing on slowness and less and less, as in the account the authors provide of

Shinichi Tsuji's *Slow Life*. This contrast motivates Hearn and Lindsay to write what may stand as motivation for this entire volume:

The pace of life and the pace of consumption are important sites of debate that cannot be ignored. In fact we suggest that the future of food will be largely influenced by debates around the future of consumption and hence our focus is on the places where people "do food": Where they eat, cook and grow.

Like several other authors in this section, Hearn and Wright distinguish "food zones" as a notion of place and scale, or in their terms—macro, meso, and micro food zones. The Food Futures Cube is finally defined as an icosikaiheptagonal geometry of three dimensions of three classes, namely "Food Activity: Eat, Cook, Grow"; "Scale of Zone: Micro, Meso, Macro"; and "Layers: Technology, People, Meanings." Of these twenty-seven pairings, Hearn and Wright offer three scenarios intended as design provocations—investigations into both positive and negative potentials: (1) a virtual reality–enabled, food-safety tracking scenario; (2) a personal food printer scenario; and (3) a scenario in which local food community practices are a matter of public policy and legislation.

10 "You Don't Have to Be a Gardener to Do Urban Agriculture": Understanding Opportunities for Designing Interactive Technologies to Support Urban Food Production

William Odom

Sustainability has emerged as a critical concern for researchers and designers of interactive systems and technologies (Blevis 2007). To date, the sustainable human-computer interaction (HCI) movement has focused heavily (but not exclusively) on environmental impacts of consumer behaviors, such as product recycling and resource consumption (see DiSalvo, Sengeres, and Brynjarsdótir 2010). These nascent works generally frame sustainability in terms of individual consumption. However, researchers are beginning to articulate the need for the HCI community to move beyond an emphasis on consumer behavior and toward an understanding of sustainability that combines environmental, social, and economic concerns (DiSalvo, Sengeres, and Brynjarsdótir 2010; Hirsch et al. 2010). In particular, exploring the needs, values, and practices of small-scale food producers has been cited as a critical point of departure to (1) foundationally consider interrelations among environmental, social, and economic sustainability, and (2) further develop HCI research and practice as an agency for sustainable ways of being (Hirsch et al. 2010). While research is emerging within contexts of developing countries and agrarian practices (e.g., Patel et al. 2010), little work exists to illustrate how the practices of small-scale urban food producers could inform the design of interactive systems aimed at supporting urban local food production.

This chapter presents findings from eight months of ethnographic fieldwork I conducted with my colleagues among members of an urban agriculture community in a large Australian city. The aim of this work is to develop a sensibility for understanding the values and practices of these communities with a critical eye toward their uses—and non-uses—of interactive systems and technologies. The findings from our fieldwork suggested several design directions aimed at helping better achieve urban agriculture community members' goals to facilitate more sustainable ways of living in the urban environment. These design directions include: developing tools and services to recode (food) waste as a strategic resource; designing services to support and extend resource exchange practices; and amplifying the visibility of urban agriculture in the city. The chapter provides an overview of related work on global food demand, food urban agriculture, and sustainable HCI. It then describes findings from our fieldwork and interviews, and concludes by discussing opportunities for designing interactive technologies and systems to support small-scale urban food production, as well as by examining implications for future research.

Background: Global Food Demand and Urban Agriculture

In 1798, Thomas Malthus speculated that global demand driven by an ever-increasing human population would surpass the ability to supply sufficient food, potentially leading to famine, war, and international political clashes (Malthus 1798/1817). According to the 2010 UN-sponsored Food and Agriculture Organization statistics, there are an estimated 925 million people living in hunger worldwide (UN World Food Program 2010). Currently, 50 percent of the global population lives in urban settings; by 2015 it is estimated that twenty-six cities worldwide will have populations of ten million people or more (Smit, Ratta, and Nasr 1996). The mass migration of populations from rural to urban areas worldwide places increased strain on local food production infrastructures, which often cannot support current demand (Smit, Ratta, and Nasr 1996).

These increasing food demands, paired with the globalization of trade, have resulted in significant consequences for social and environmental sustainability (Nutzenadel and Trentmann 2008; Stuart 2009). On one hand, the average store-bought produce item is estimated to travel between 1,500 and 2,500 miles before being purchased for consumption, resulting in significant levels of pollution and carbon dioxide emissions (Pretty et al. 2005). On the other hand, general lack of availability of fresh produce has been cited as a key contributor to increased consumption of processed foods, which have been linked to the growing epidemic of diabetes and obesity in the Western world (Smit, Ratta, and Nasr 1996). Additionally, the movement of food from its origin to distant markets contributes to significant food waste; previous research has estimated European and American food distributors, supermarkets, and consumers throw away between 30 to 50 percent of their food supply due to it perishing (Stuart 2009).

Urban agriculture is the practice of integrating low-energy food production techniques within city boundaries to increase the amount of fresh food available to urban consumers (Vijoen 2005). As cities expand to accommodate growing populations, they often occupy significant amounts of agricultural land. Urban agriculture has the potential to bring this land back into productive use to meet current and future produce demands, in addition to stimulating local economies through the sale of locally grown food.

(Re)connecting Local Food Producers and Consumers

Although there exists wide consensus that current food supply arrangements are in need of critical assessment, the issue of how to equitably and effectively approach addressing the global food demand is complex; many researchers have argued it goes beyond understanding the problem solely as an empirical matter of supply and demand (e.g., Fry 2009; Stuart 2009). Recent work has speculated on the benefits of connecting the local practices, process, and cultures of food production and consumption in the service of (re)shaping human perceptions of local produce to support more environmentally sustainable behaviors (Fry 2009; Kneafsey et al. 2008).

However, it is important to understand the reframing of consumption within this context. In contrast to consumable products, Tony Fry (2009) highlights the opportunity local urban food production systems provide to support metabolic forms of consumption in which residual (or unconsumed) elements from food production, preparation, and consumption can be used—in essence "metabolized"—to fuel future cycles of production (e.g., through creating nutrient-rich compost). Additionally, researchers have argued that things such as urban waste or idle land and water bodies ought to be recognized collectively (and used) as key resources to support and expand agricultural practices throughout cities (Smit and Nasr 1992). Local political structures, however, and, in some cases, commercial interests, have been shown to complicate the reuse of these often under-considered, yet largely prevalent urban resources (Hinrichs 2003). Although there appear to be opportunities in more explicitly connecting the citizens, stakeholders, resources, and spaces involved in food production (and consumption), little work has explored the role interactive technologies might play in supporting—and potentially expanding—these processes within urban settings.

Sustainable HCI and Small-Scale Food Production

Issues related to food and food production have been steadily gaining purchase within the HCI community. Researchers have explored the role sensors could play in reducing transport of perishable food (Ilic, Staake, and Fleisch 2009) and supporting large-scale agricultural production (Burrell, Brooke, and Beckwith 2004). Others have investigated how educational technology interventions could facilitate learning agricultural practices (e.g., Pearce, Smith, and Murphy 2009). Andrea Grimes and Richard Harper (2008) call for HCI research and practice to more fully embrace the social, celebratory aspects of food in everyday life. Catalina Davis and colleagues (2009) suggest several possibilities for social and mobile computing to support farming in developing regions.

Very recently, a number of HCI researchers (Blevis and Morse 2009; Brynjarsdóttir and Sengers 2009; Choi and Blevis 2010; Choi et al. 2009; DiSalvo et al. 2010; Hirsch et al. 2010; Patel et al. 2010), have collectively articulated the virtues of exploring the intersection of HCI and small-scale food production and examine how work in this area could shape future sustainable HCI research. More generally, prior research has illustrated how exploring the values, practices, uses—and non-uses (Satchell and Dourish 2009)—of groups existing outside of what could be considered "mainstream" user populations can productively inform new areas of inquiry within the HCI community (e.g., Woodruff et al. 2008). Our study further builds on these emerging areas through describing the values and practices of participants in an urban agriculture community, as well as key opportunities they suggest for designing interactive systems to support small-scale urban food production.

Methods

Our study was conducted in a large Australian city. Through personal contacts we established rapport with an urban agriculture community, which primarily operated at two community garden sites in close proximity. These sites were situated in a prominent urban area and offered opportunities to understand key ways in which urban agriculture practice fit within broader urban processes and culture (as well as emergent barriers and complications).

We conducted fieldwork to gain a better understanding of community members' practices and values, as well as to record in situ insights into the everyday interactions and collaborations associated with maintaining the community-oriented urban agriculture sites. Additionally, we observed members perform a variety of other activities around the city. Throughout these observations, we also engaged everyday citizens from a range of backgrounds who attended public workshops and demonstrations.

We also conducted semistructured interviews with nineteen community members across both urban agriculture sites. These interviews explored such issues as motivations for participation, individual and group gardening practices, perceptions of urban space, uses and non-uses of technology, and perspectives on potentially effective or ineffective technology interventions. This portion of the study produced rich data consisting of handwritten field notes, audio recordings, and photographs. We listened to recordings and transcribed relevant segments. We repeatedly reviewed these relevant portions and organized them into themes; we then coded the textual and visual documents using these emergent themes.

Findings

Over the course of our fieldwork, we routinely observed members of a local urban agriculture community, which operated at two collocated community gardens. It was estimated the community comprised forty-six members. Participants reported a range of occupations including botanist, schoolteacher, musician, health-food store clerk, computer technician, textile designer, jeweler, and unemployed. While the occupations of participants and motivations to join the community (described below) varied, it was clear that members in general regarded their values, practices, and culture to be alternative to other groups in the city.

Members largely worked together at the sites. Generally half of the produce yield was divided among participating members, while the rest was sold or exchanged for resources to maintain the respective sites. Members were required to maintain a certain set of biweekly hours to receive produce. Work to fulfill these hours consisted of direct interactions with the site (e.g., upkeep of gardens) as well as a range of tasks associated with obtaining the necessary resources to sustain the garden (e.g., organic composting materials, various tools). In some cases, work also consisted of scoping urban areas for potential future sites, holding public workshops on urban agriculture and "guerrilla gardening" techniques, and organizing public demonstrations.

Participants reported a range of motivations for becoming community garden members. These included learning new gardening skills, engaging in productive community action, desiring to be closely linked with food production and consumption, and supporting food localization.

An early and consistent theme across our interviews centered on how participants' engagement with personal and community gardening played a role in shaping their perceptions of food. For example, Mindy, a participant who had been a member of the community for six months, describes this transition: "Once I started here, I became more conscious of [food]. I'd try to never let it rot, right, you just try not to. I'd make sure someone'd use it if I couldn't." It is important to point out, however, that despite such orientations toward food, several participants also reported at times neglecting personal plants by leaving them under-watered or letting food over-ripen. These instances were nearly always attributed to being occupied with other routine or unexpected tasks. Nonetheless, what we want to draw attention to is how engagement with the gardening site and the yielded produce tended to shape how participants perceived and related to food. As we will later describe, somewhat comparable shifts in perspective also emerged for other citizens not directly involved in community gardens. In what follows, we describe key community virtues as well as perceptions and uses of technology. We then describe several key ways in which technology was drawn on to coordinate or project the work and practice of community members in agriculture sites and other urban spaces.

Resourcefulness as a Core Community Virtue

Intentionally acting and living resourcefully appeared to be a pervasive value shared across community members. For example, our participants widely valued the premise that urban agricultural practice is fundamentally about the appropriation of urban space for productive re-use. Additionally, they valued drawing on materials that would otherwise be treated as waste and creatively putting them to beneficial use. We observed many instances in which discarded materials such as barrels, aluminum cans, and corrugated iron roofing segments were composed into new assemblies for a variety of uses, ranging from rooftop rainwater collection systems to ground-level composting receptacles. This theme also emerged through members' treatment of leftover organic materials to produce garden fertilizer: "The problem is people don't know most of what they throw out can be used. They think that it's trash, so they chuck it in the bin. Our mindset is different. We use what people throw out and what's leftover from our meals and [garden] work to grow more food" (Liam). Liam's reflection is exemplary of how leftover food "waste" was drawn on—and recognized as—a resource directly metabolized by the local site's ecosystem to produce new crops; this process was described by many members as an embodiment of the community's resourceful nature and practice.

Resistance to Technological Augmentation of Gardening

During each interview, after participants had described their everyday activities related to urban agriculture in and around the city, they were asked to reflect on how new technology might be used to positively support their practices. Often participants struggled at first to come up with ways in which they would use technology; when discussions turned to how it might be applied within a gardening context, potential interventions were nearly always regarded as highly contentious.

For example, the notion of using sensing technology to provide dynamic information on soil moisture levels or chemical compositions was regarded as one that might produce a series of potential negative consequences. Several participants described how ongoing direct interaction with the site (e.g., garden plots, plant installations, soil) helped establish a reflective sensibility for understanding, predicting, and reacting to how environmental changes would affect the local sites. It was widely reported that such a sensing system could be relied on too heavily, subverting members' ongoing development of environmental knowledge and intuition. Additionally, instances when newcomers to the community had questions during garden work sessions were viewed as key opportunities for more experienced members to offer insights into urban agriculture practice and techniques. These situations were regarded as crucial to strengthen community member relations through social interaction and instruction. Some senior participants speculated that the presence of information from this kind of sensing system could supplant didactic interactions between new and more experienced members, thus potentially complicating the development of community relationships and practice. While sensing technology could potentially make it "easier" to collect garden plot–level information, participants speculated that it could potentially undermine social practices catalyzing transfer of knowledge, and, potentially, the community's sustainment.

On the whole, it was clear that the application of new technology to directly augment in-garden practices did not map to community values nor appear to be beneficial. However, members did appropriate interactive systems for several purposes. In the follow section we describe how technology use was often intertwined with material and social practices associated with the work of (1) managing and/or exchanging resources to maintain garden sites, as well as (2) documenting political action and gardening practice to project the agenda of urban agriculturalists within the broader public sphere.

Maintaining Community Gardens: Emergent Systems of Exchange and Shared Use

One of the biggest sources of work for community members centered on obtaining and managing a constant flow of tools, organic materials, and harvested food into and out of gardening sites. Mobile phones were heavily relied on to coordinate action and move resources between sites, as well as obtain resources from various urban locations. One major set of activities had to do with obtaining a steady flow of organic materials to create nutrient-rich compost for the gardens. Several members routinely ventured around the city to monitor dumpsters of local restaurants and other businesses known to discard compostable materials. When members spotted discarded caches of usable organic materials, text messages detailing the location and contents were broadcast to the community to solicit immediate help in collection and transport. These actions straddled a fuzzy boundary of social and legal acceptability, which had caused clashes with the police.

A large portion of community members had since individually turned to citizens in their local neighborhoods to acquire resources to maintain garden sites, which included soil, caches of compostable food materials, and various tools. Differing types of compost material, as well as weather events (e.g., thunderstorms, heavy wind) and seasonal weather patterns required a range of tools and resources to cater to the sites' changing conditions. Participants described several ways in which they acquired tools and materials. Most commonly, these practices began with door-to-door visits to citizens' homes in their respective neighborhoods. After personal trust and rapport had been established, many members reported migrating these practices to small individually run email listservs of citizens in their respective neighborhoods to send out and coordinate requests for various resources. In some cases, participants described how requests through listservs became more successful after they began including small amounts of locally grown produce in exchange for materials or temporary use of tools.

Although email listservs have long been used to coordinate activities among groups, we want to draw attention to how this technology was leveraged to support informal systems of local barter and shared use among community members and various citizens. Irwin, along with several other participants, described positive outcomes engendered through engagement with these systems over time. On one hand, participants reported how interactions through the listserv catalyzed several citizens' interest in gardening, which led to the creation of their own home plots or potted plants. In other instances, they provided an opportunity for citizens that either lacked resources to have gardens, or were simply uninterested in growing their own food, to participate in local food production and consumption cycles.

Over the course of our study we shadowed several participants on trips to collect various resources from neighbors on a local listserv. We accompanied Samuel, a senior community member, on one trip where we had the opportunity to interview Julian, a middle-aged citizen participating in the email listserv for over a year, primarily to donate composting materials. Julian described how engaging with the gardening community through the listserv had changed his attitude toward local food production and use of public space: "I don't have time to garden myself. [Samuel] showed me how I could do something by saving stuff, like coffee [grounds], egg shells . . . all I do is put 'em in a different bin, easy enough. . . . I think more about what I chuck out. This isn't trash (motions to compost bin). I know where it's going—in the ground helping grow more food." Julian further reflected on broader issues he had begun to consider since participating on the listserv: "I started thinking about [my] land too. . . . As I say, I don't have the time to garden in my [backyard], but someone else could put some of it to use. . . . Why aren't we using parts of public parks for it too?"

We also encountered citizens who exchanged temporary use of their tools for small amounts of produce via their local listserv. For example, Jane, a mid-fifties banker and woodworker hobbyist, described how this type of exchange shaped the way she conceptualized participation within urban agriculture: "I'm too old to spend all day outside and really [gardening] wasn't ever an interest of mine. . . . But I have all these tools and it's satisfying to see they're getting put to use." Ed, a mid-thirties coffee shop owner, further reflected on how loaning tools shaped his thoughts on shared use and ownership: It "made me think about all the stuff I have and how little I use most of it now. . . . I use some of these tools once or twice a year. . . . I'm glad to trade them for a while and get back some of the food and be part of that process." However, we did encounter some instances in which citizens possessed useful tools, but elected to only exchange composting materials due to anxiety over tools going missing or unreturned: "My tools and kit, they can't go. I don't know if I'll get 'em back. . . . I mean I trust [the community] but you can lose'em easily. You know?" (Harold, mid-sixties, retired construction foreman).

Nonetheless the study found that whether through exchanging resources such as tools or leftover composting materials, we commonly found these practices shaped how citizens thought about the food they received as well as their participation in the process as a whole: "Even though I'm not doing it [gardening], knowing where the food comes from and contributing to it changes how you think about it" (Sarah, late twenties, school teacher). When reflecting on field notes documenting this emergent theme with Samuel after returning from a collection run to a community garden site, he mentioned, "One of the biggest misunderstandings about the whole thing, is you don't have to be a gardener to do urban ag[riculture]. . . . It's about bringing together different people doing different things to make it work." Samuel's reflection raises a salient point: the "doing" of urban agriculture extends beyond gardening itself to systemically managing and coordinating diverse streams of contributions from various members, citizens, and stakeholders. We found participation in this process (largely through email listservs to coordinate face-to-face interactions) appeared to shape citizens' perceptions of reusable organic materials and "waste," how urban land could be used, and issues of ownership and exchange. Despite these emergent benefits, the amount of citizens participating in these local neighborhood listservs remained relatively low on a citywide scale. As community members sought to expand to new sites for community gardens, additional resources (and members) were first required. We next describe tactics employed by participants to project their agenda into the public sphere in the service of engaging citizens and stakeholders in urban agriculture issues and practices.

Re-envisaging Urban Land and Stimulating Public Interest in Urban Agriculture

We observed a deep tension between community members' perspectives on how public urban land ought to be used for food production and current local government policies preventing gardening on public land outside of a few designated areas. Community members held several events in which public spaces, such as sidewalks, parking spaces, and public parks, were temporarily appropriated and populated with small installations of food-producing plants to re-image how city space could be used. During these demonstrations, members directly engaged citizens passing by, presenting them with information about urban agriculture. These events were documented and integrated into ongoing Web campaigns run by members aimed at envisaging future scenarios of urban life in spaces around the city. However, these demonstrations disrupted patterns of everyday urban life in local areas (e.g., restricting traffic flow, sidewalk space). Consequently, obtaining permits for future demonstrations from the city council became difficult.

As a result, some community members had adopted more radical practices aimed at marking sites ripe for agricultural interventions and stimulating public concern over how urban land ought to be used. These "guerrilla gardening" workshops most commonly catalyzed initial actions in support of these objectives; they were open to the public and typically included a mix of community members, activists, and, to a smaller extent, interested citizens. The goals of workshops were to identify urban land or infrastructure that could be used for food production, and to develop interventions targeting these sites. Workshop organizers considered it fundamentally important that food-producing plants function as their primary medium of expression: "Planting something is political . . . but it's different than putting up fliers . . . it's not saying 'consider how this land could be used.' It says 'this land should be used, and this is a sign of it happening'" (Zack, workshop organizer).

We observed a range of techniques taught at workshops to construct interventions. These included lessons on how to optimize group processes associated with installing small installations on public property (typically at nighttime), to making and deploying "seed bombs"—compressed assemblies of soil, seeds, and live vegetation encompassed in biodegradable containers that could be thrown onto land, often over fences or other infrastructural barriers. At the conclusion of workshops, plans were typically made to disseminate interventions at sites in the city and, at times, revisit areas in which plants had begun to grow to draw further attention to unused land.

We accompanied participants on three interventions. A particularly compelling demonstration unfolded against the backdrop of a vacant fenced-in urban lot owned by the city, which participants reported had gone unused for nearly two years. A few months previously, workshop members deployed a large assortment of seed bombs on the lot and several small plants had begun to emerge. During our observation of the demonstration, several citizens passing by stopped when organizers approached them to discuss the site and current city policy relegating land use. After initial engagement with workshop members, we conducted quick informal interviews with several citizens, the majority of which lived or worked locally in the area. For example, Margery, resident of a local apartment complex, reflected on the unanticipated land use: "Didn't think it'd be a garden, but you know makes me think maybe we're better off with it than another parking lot." What we want to highlight is how the presence of plants within unused urban land challenged conventions and introduced new possibilities of how it could be used. These instances were also documented by workshop members and distributed though locally run urban agriculture and guerrilla gardening websites.

In addition to documenting potential for future urban agricultural sites, one of the most prevalent uses of the Internet by community members centered on documenting key areas in which urban agriculture was already taking place. With the exception of a limited number of community gardens around the city, most urban agriculture takes place on rooftops, backyards, and other areas largely out of sight to citizens at large. The aim of these community-run websites was to highlight key places around the city in which these practices were unfolding in order to generate additional exposure and public interest. Despite these collective efforts, the majority of members reported dissatisfaction over the inability to acquire deeper purchase among city citizens and stakeholders, and gain political traction.

Design Implications and Opportunities

Urban agriculture and its place in the city is complex. A key contribution of this study is to present and interpret observations of urban agriculture community members and to describe barriers faced by members as they attempted to expand their practice in the city. These findings have a number of implications for the design of interactive systems aimed at supporting small-scale food production. The following sections outline three design implications with accompanying opportunities to help guide future research in the HCI community.

Developing Tools and Services to Recode (Food) Waste as a Strategic Resource

A considerable amount of work undertaken by community members centered on obtaining necessary resources to sustain sites and, in doing so, attempting to persuade citizens and stakeholders that resources considered as "waste" could be productively used. Garden sites in dense urban areas can be in close proximity to restaurants in which compostable food waste is frequently discarded. Nonetheless, there were clear tensions between community members and several restaurant owners that complicated exchange of these resources.

This suggests a viable opportunity for designing interactive tools and services that might better facilitate relationships between local urban agriculture sites and commercial restaurants producing compostable organic "waste." Restaurant food waste could be metabolized within local urban agriculture sites, which in turn, could produce fresh food catering to restaurant menus. Ongoing exchanges engendered through these systems could demonstrate how food waste can be used—and perhaps symbolically recoded (Fry 2009)—as a strategic resource, potentially catalyzing future local food production and (metabolic) consumption. These services could be paired with interactive media campaigns aimed at leveraging leftover organic materials as boundary objects to mediate conversations between members of local community gardens and business owners and to broker future exchanges.

Additionally, long-term uses of food producer–consumer services could provide opportunities for establishing case studies of the benefits produced when urban agriculture initiatives are systemically integrated within local social and economic structures. We imagine these services could be expanded to peri-urban farms in further support of local cycles of exchange, production, and consumption. This direction could complement emerging work exploring the role mobile technologies and platforms could play in supporting agricultural practices among peri-urban and rural dwellers (Davis et al. 2009). More broadly, future research investigating the local social and cultural factors required to symbolically recode "waste" materials as strategic resources could build on existing work in (Strengers 2008) and outside (e.g., Chappels and Shove 1999; Fry 2009) of HCI, exploring how everyday interactions might be reshaped in the service of promoting more sustainable ways of being.

Designing Interactive Systems to Support and Extend Resource Exchange Practices

From continually acquiring composting materials to obtaining an evolving set of tools due to changing environmental conditions, a considerable amount of work to sustain community gardens unfolded outside of the sites themselves. Emergent informal systems of sharing and exchange appeared to provide valuable opportunities for citizens to support local food production at community gardens, without directly being involved with work at those sites.

These findings suggest several possible ways to engage citizens with local food production communities. In particular, there appears to be a major opportunity in designing interactive systems to better support the exchange of resources between individual citizens and urban gardening sites. These systems could more explicitly make public the evolving needs of particular urban sites and the types of resources offered by citizens. We also found some citizens resisted sharing tools due to no clear way of knowing the whereabouts and person(s) accountable for them; interactive technologies could be leveraged within these systems in the future to better keep track of various tools' locations and histories of use. More broadly, systems designed to support this direction could aim to connect, communicate (and perhaps provide a platform to moderate), the diverse and evolving community member, citizen, and stakeholder needs, political issues and social connections that characterize the work of urban agriculture. Importantly, these digital sites could highlight contributions from individual citizens and illustrate how their contributions work to sustain collective action. This direction could build on recent work calling for sustainable HCI research to extend focus to designing technologies that link people through their actions in the service of organizing for social change (e.g., Aoki et al. 2009; Dourish 2010).

More generally, as programs of sustainable HCI research continue to expand, there seems to be a large opportunity to explore how services could be designed to encourage shared use of equipment and resources among diverse groups in—and outside of—urban settings. Issues of sharing and ownership are clearly shaped by social, cultural, economic, and political structures attendant to a local setting. Clearly more research is needed on how systems facilitating exchange and shared use may (or may not) appropriately fit and function within local ecologies of use. Emerging research on opportunities and barriers to shared use in Western (Goodman 2010) and non-Western (Ito et al. 2010) settings could be leveraged in the service of this direction, as could the wealth of prior work at the intersection of service design, dematerialization and social change (e.g., Manzini 1999; Tonkenwise 2005).

Amplifying Visibility of Agriculture in and on the City

A key issue inhibiting the broader uptake of urban agriculture practices by city citizens appeared to be a lack of visibility of urban agriculture sites; the physical places and the political issues implicated in them were largely outside of public sight. This suggests significant opportunities for designing systems that amplify the presence of urban agricultural practices and sites on, in, and around the city. We imagine systems could be designed to implicitly or explicitly code urban structures, places, and land that use (or offer) space for local food production. For example, interactive maps, geo-tags, and citizen-enabled content could be applied to convey where urban agriculture practices are unfolding around the city, how and where produce is consumed and sold (and metabolized), and key stakeholders participating in these initiatives. There also appear to be opportunities in developing interactive media facades that could be projected on urban spaces to envisage viable ways in which key sites could be brought into productive use in the future. These systems could link public space and materials with critical commentary on political issues, such as public land use. Such applications could collectively draw on prior work exploring efforts to re-image urban settings (DiSalvo and Vertesi 2007), use public displays to increase a citizen's knowledge of local urban settings (Dalsgaard and Halskov 2010), and project a public rhetoric reflective of the public's concerns and agendas (DiSalvo et al. 2008).

We also found the act of planting food-producing plants in underutilized areas was considered a unique political act of reclaiming public space and stimulating reflection on how it could be used. We imagine there are opportunities to extend this form of grassroots public authorship through robotic and mobile technologies, which, for example, could enable citizens to spread seeds throughout the city and document these actions by making them publicly viewable through the Internet, public displays, and perhaps even through images re-projected onto the cityscape itself. Previous work has explored approaches to designing with and for political activists (Hirsch 2009) as well as the role low-cost robots can play in traversing urban landscapes to project underrepresented groups' viewpoints (Kuznetsov, Paulos, and Gross 2010). Such research can be leveraged to extend the use of interactive technology for public "authorship" of urban places and to support the political agendas of urban agriculture communities.

Conclusions and Future Work

This study presents work contributing to the emerging intersection of HCI and sustainable food production. The aim of this study is to illustrate one way in which the values and practices of urban agriculture community members could inform the design of interactive systems aimed at supporting food production in the city. We proposed a number of design opportunities, which collectively emphasized a broader focus on drawing key resources, stakeholders, and social connections and issues into systemic relation, as opposed to device-level interventions augmenting specific gardening practices. However, we can imagine garden-level sensing interventions could be useful in a variety of other situations, such as when less community support is available and remote farming applications are required. Clearly more research is needed in this area. However, how the "local" is constructed can vary significantly across sociocultural, economic, political, and geographic boundaries (Hinrichs 2003). As such, how it is characterized and constituted must be considered in the design of future systems embedded in diverse local contexts.

Finally, it is critical to acknowledge that there is nothing inherently sustainable about the process of localizing food systems, which operate in the extent to which those in power want them, or allow them, to (Feagan 2007). Comparative studies are needed to better understand how different communities organize themselves on a community level and in the city, the ways in which their political structure and power distribution shape these processes, and, ultimately, how this knowledge could lead to the design of more effective and extensible systems supporting the localization of food production.

Acknowledgments

I am indebted to Tony Fry, Anne Marie Willis, and Eli Blevis for their guidance and continued production of inspiring work. I would also like to thank Nora Kinnunen, Petra Perolini, and Coral Gillett for their scholarship, collaboration, and willingness to grapple with new ideas. This work was partly supported by a Fulbright fellowship.

SPAGHETTI SQUASH WITH CRANBERRY AND SAGE

Makes 6 to 8 servings as a side dish
The squash and sage in this recipe can easily be grown in one's garden (or found at a local farmers' market). Be careful to harvest spaghetti squash when it is fully mature and leave two inches of the vine when it is cut. This recipe was invented by Sam Shoulders. This dish has a long tradition of being consumed at dinner parties laden with wild speculation, deep meditation, and good conversation. This is no mere coincidence, so beware!

Ingredients

1 spaghetti squash

1–1¼ cup of dried cranberries

1 cup of port (substitute ½ cup cranberry juice and ½ cup water if desired)

5 sage leaves (more or less to taste)

2 tbsp. butter

1 tbsp. olive oil

Salt and pepper to taste

Preparation

Cooking the Spaghetti Squash

1. Preheat the oven to 375°F. Prick one medium-to-large spaghetti squash with a skewer (or the tip of a sharp, thin-bladed knife) so it won't burst while cooking. Place in a shallow pan and bake for one hour.

2. When it's cool enough to handle, cut it in half lengthwise with a serrated knife. Scoop out the seeds and loose fibrous material in the center with a spoon, and discard. Take a fork and gently scrape along the edges with the tines, moving toward the interior of the squash, until only the skin remains; it should break apart in strands that look like spaghetti.

3. If you scrape carefully you should be able to use one of the squash skins as a serving dish, allowing a portion of the squash to spill over onto a platter underneath.

Port and Dried Cranberries Sauce

1. Pour the cup of port in a sauté pan. Turn heat on low and gently increase the heat until the port simmers.

2. Add the dried cranberries and cook for 10 minutes (always over low heat).

Sage Brown Butter

1. Take fresh sage leaves and stack them on top of each other. Then take a sharp knife and slice the stack into thin slivers.

2. Place one tbsp. of butter in another sauté pan. Turn the burner up to medium-high until the butter melts. Place the slivers of sage in the hot butter. The butter should begin to brown, and when it does add the remaining tbsp. of butter. Remove when the butter develops a light brownish color.

Bringing It All Together

1. Place the spaghetti squash in a large sauté pan or stir-fry pan over low heat, pour the sage brown butter over it, and begin to toss it until just mixed in. Add sea salt and pepper to taste, just enough to add some depth to savory aspect of the dish.

2. When heated through, spoon the squash into a serving bowl (or into your clean spaghetti-squash skin, placed on a platter to catch the overflow). Now take the port sauce and pour slowly over the squash. Garnish the squash with a sprinkle of dried cranberries if desired.

References

Aoki, Paul. R. J. Honicky, Alan Mainwaring, Chris Myers, Eric Paulos, and Allison Woodruff. 2009. A vehicle for research: using street sweepers to explore the landscape of environmental community action. In *Proceedings of the 27th International Conference on Human Factors in Computing Systems (CHI '09)*, 375–384. New York: ACM.

Battaile, Georges. 1988. *The accursed share: An essay on general economy*. New York: Zone Books.

Blevis, Eli. 2007. Sustainable interaction design: invention & disposal, renewal & reuse. In *Proceedings of the 5th International Conference on Human Factors in Computing Systems (CHI '07)*, 503–512. New York: ACM.

Blevis, Eli, and Susan Morse. 2009. Sustainably ours: Food, dude. *interactions* 16 (2): 58–62.

Born, Branden, and Mark Purcell. 2006. Avoiding the local trap: Scale and food systems in planning research. *Journal of Planning Education and Research* 26:195–207.

Brynjarsdótir, Hronn, and Phoebe Sengers. (2009). Ubicomp from the edge of the North Atlantic: Lessons from fishing villages in Iceland and Newfoundland. In *Workshop proceedings of GlobiComp, UBICOMP '09*.

Burrell, Jenna, Tim Brooke, and Richard Beckwith. 2004. Vineyard computing: Sensor networks in agricultural production. *IEEE Pervasive Computing / IEEE Computer Society [and] IEEE Communications Society* 3 (1):38–45.

Chappels, Heather, and Elizabeth Shove. 1999. The dustbin: a study of domestic waste, household practices and utility services. *International Planning Studies* 4 (2):267–280.

Choi, Jaz Hee-jeong, and Eli Blevis. 2010. HCI and sustainable food culture: A design framework for engagement. In *Proceedings of the 6th Nordic Conference on Human-Computer Interaction: Extending Boundaries*, 113–117. Reykjavik, Iceland: ACM.

Choi, Jaz Hee-jeong, Marcus Foth, Greg Hearn, Eli Blevis, and Tad Hirsch. (2009). Hungry 24/7? HCI design for sustainable food culture. In *Proceedings of the 21st Australasian computer-human interaction conference, OZCHI 2009: Open 24/7*. New York: ACM.

Dalsgaard, Peter, and Kim Halskov. 2010. Designing urban media facades: Cases and challenges. In *Proceedings of the 28th International Conference on Human Factors in Computing Systems (CHI '10)*, 2277–2286 New York: ACM.

Davis, Catalina, Mark Bailey, Jime Christensen, Jason Ellis, Thomas Erickson, Robert Farrell, and Wendy A. Kellogg. 2009. Mobile applications for the next billions: A social computing perspective. In *Proceedings of the 28th international conference on human factors in computing systems (CHI '09)*, 4745–4750. New York: ACM.

DiSalvo, Carl, Illah Nourbakhsh, David Holstius, Ayca Akin, and Marty Louw. 2008. The neighborhood networks project: A case study of critical engagement and creative expression through participatory design. In *Proceedings of the tenth conference on participatory design (PDC '08)*, 41–50. New York: ACM.

DiSalvo, Carl, Phoebe Sengers, and Hronn Brynjarsdótir. 2010. Mapping the landscape of sustainable HCI. In *Proceedings of the 28th International Conference on Human Factors in Computing Systems (CHI '10)*, 1975–1984. New York: ACM.

DiSalvo, Carl, Ann Light, Tad Hirsch, Christopher A. Le Dantec, Liz Goodman, and Katie Hill. 2010. HCI, communities and politics. In *CHI EA '10 Extended Abstracts on Human Factors in Computing Systems*, 3151–3154. New York: ACM.

DiSalvo, Carl, and Janet Vertesi. 2007. Imaging the city: Exploring the practices and technologies of representing the urban environment in HCI. In *CHI EA '07 Extended Abstracts on Human Factors in Computing Systems*, 2829–2832. New York: ACM.

Dourish, Paul. 2010. HCI and environmental sustainability: The politics of design and the design of politics. In *Proceedings of the 8th ACM Conference on Designing Interactive Systems (DIS '10)*, 1–10. New York: ACM.

Feagan, Robert. 2007. The place of food: Mapping out the "local" in local food systems. *Progress in Human Geography* 31 (1):23–42.

Fry, Tony. 2009. *Design futuring: Sustainability, ethics and new practice.* Oxford: Berg.

Grimes, Andrea, and Richard Harper. 2008. Celebratory technology: New directions for food research in HCI. In *Proceedings of the Twenty-Sixth Annual SIGCHI Conference on Human Factors in Computing Systems(CHI '08)*, 467–476. New York: ACM.

Hinrichs, C. Clare. 2003. The practice and politics of food system localization. *Journal of Rural Studies* 19 (1):33–45.

Hirsch, Tad, Phoebe Sengers, Eli Blevis, Richard Beckwith, and Tapan Parikh. 2010. Making food, producing sustainability. In *CHI EA '10 Extended Abstracts on Human Factors in Computing Systems*, 3147–3150. New York: ACM.

Hirsch, Tad. 2009. Learning from activists: Lessons for designers. *Interactions* 16 (3):31–33.

Ilic, Alexander, Thorsten Staake, and Elgar Fleisch. 2009. Using sensor information to reduce the carbon footprint of perishable goods. *IEEE Pervasive Computing / IEEE Computer Society [and] IEEE Communications Society* 8 (1):22–29.

Ito, Erika, Mikiko Iwakuma, Shunsuke Taura, Tomoaki Hashima, Yu Ebihara, and Naohito Okude. 2010. Sharebee: Encouraging *osusowake* to promote community development. In *Proceedings of the 8th ACM Conference on Designing Interactive Systems (DIS '10)*, 228–231. New York: ACM.

Kneafsey, Moya, Rosie Cox, Lewis Holloway, Elizabeth Dowler, Laura Venn, and Helena Tuomainen. 2008. *Reconnection producers, consumers and food: Exploring alternatives.* Oxford: Berg.

Kuznetsov, Stacey, Eric Paulos, and Mark D. Gross. 2010. Wallbots: Interactive wall-crawling robots in the hands of public artists and political activists. In *Proceedings of the 8th ACM Conference on Designing Interactive Systems (DIS '10)*, 208–217. New York: ACM.

Malthus, Thomas Robert. 1798/1817. *Essay on the principle of population as it affects the future improvement of society.* London: Clowes.

Manzini, Ezio. 1999. Strategic design for sustainability: Towards a new mix of products and services. [IEEE.] *Proceedings of EcoDesign* 99:434–437.

Nutzenadel, Alexander, and Frank Trentmann, eds. 2008. *Food and globalization: Consumption, markets and politics in the modern world*. Oxford: Berg.

Odom, William. 2010. "Mate, we don't need a chip to tell us the soil's dry": Opportunities for designing interactive systems to support urban food production. In *Proceedings of the 8th ACM Conference on Designing Interactive Systems* (DIS '10), 232–235 New York: ACM.

Patel, Neil, Deepti Chittamuru, Anupam Jain, Paresh Dave, and Tapan S. Parikh. 2010. Avaaj Otalo: A field study of an interactive voice forum for small farmers in rural India. In *Proceedings of the 28th international conference on Human factors in computing systems (CHI '10)*, 733–742. New York: ACM.

Pearce, Jon, Wally Smith, and Bjorn Naohn Murphy. 2009. SmartGardenWatering: Experiences of using a garden watering simulation. In *Proceedings of the 21st Australasian Computer-Human Interaction Conference, OZCHI 2009: Open 24/7*, 217–224. New York: ACM.

Pretty, J., A. Ball, T. Lang, and J. Morison. 2005. Farm costs and food miles. *Food Policy* 30 (1):1–20.

Satchell, Christine, and Paul Dourish. 2009. Beyond the user: Use and non-use in HCI. In *Proceedings of the 21st Australasian Computer-Human Interaction Conference, OZCHI 2009: Open 24/7*, 9–16. New York: ACM.

Smit, Jac, Annu Ratta, and Joe Nasr. 1996. *Urban agriculture: Food, jobs, and sustainable cities*. New York: United Nations Development Programme (UNDP).

Smit, Jac, and Joe Nasr. 1992. Urban agriculture for sustainable cities: using wastes and idle land and water bodies as resources. *Environment and Urbanization* 4 (2):141–152.

Strengers, Yolande. 2008. Smart metering demand management programs: challenging the comfort and cleanliness habitus of households. In *Proceedings of the 20th Australasian Conference on Computer-Human Interaction (OZCHI '08)*, 9–16. New York: ASM.

Stuart, Tristram. 2009. *Waste: Uncovering the global food scandal*. New York: W. W. Norton & Company.

Tonkinwise, Cameron. 2005. Is design finished? Dematerialisation and changing things. *Design Philosophy Papers* 2:20–30.

Vijoen, Andre. 2005. *Continuous productive urban landscapes*. Burlington, MA: Architectural Press.

United Nations World Food Program. 2011. Hunger: Who are the hungry? http://www.wfp.org/hunger/who-are.

Woodruff, Alison, Jay Hasbrouck, and Sally Augustin. 2008. A bright green perspective on sustainable choices. In *Proceedings of the Twenty-Sixth Annual SIGCHI Conference on Human Factors in Computing Systems (CHI '08)*, 313–322. New York: ASM.

11 Augmented Agriculture, Algorithms, Aerospace, and Alimentary Architectures

Jordan Geiger

The increasingly prevalent relations between food production, human-computer inter-action, and land use planning differ by scale and geographic location, among other factors. Each of these exert force on food systems, often with conflicting goals, byprod-ucts for public health, and little input from ordinary consumers. How can we confront the vitally important and yet increasingly complex tangle of these shifting forces, and how can design practices contribute to organizing (even loosening) this tangle to produce new, more healthful orders? While the general public is largely unaware of the extent of the role of computing and electronics, trade journals such as *Computers and Electronics in Agriculture*[1] are testament to emergent but already highly advanced industrial agriculture methods and fields of research, including, for example, Compu-tational Intelligence in Crop Production, Applications of Artificial Neural Networks and Genetic Algorithms, and Agriculture Systems and Global Positioning Systems in Agriculture. While these fields clearly concern regional, landscape, and even urban and architectural design, they have remained largely absent in the discussions, devel-opments, and implementation of the built environment. Yet design of the built envi-ronment at all these scales is directly concerned with forms of interaction design (between human actors and between humans and their spaces), which in turn includes purview over human-computer interaction (HCI) and the roles of life resources like food in the health of cities. As architects today are growing more and more engaged in an awareness of this purview, their skill sets are beginning to expand and draw on research into HCI, food systems, and more. These shifts are giving rise to new fields of design practice such as landscape urbanism (Waldheim 2006) and situated technolo-gies (Greenfield and Shepard 2007), and the opportunity for an engagement and a meaningful contribution from the fields of design. These areas of design research and practice are themselves hybrids of sorts that reflect both new threats and new oppor-tunities within the design professions, where practice is responding to industrial and cultural changes that indicate a decline in demand for the type of fee-for-service work that has previously been the basis of most firms' work; simultaneously, these areas reflect a need for complex research and diversified forms of expertise that include urban design, legal skills, computer programming, and design of microelectronics. Within this context, the convergence of HCI, food, and architecture is yet more salient as a sort of hybrid of these hybrids. Seen this way, a subject such as this appears not peripheral to the concerns and practices of architecture in both the academy and in practice, but rather central to their transformations.

This chapter identifies some surprising but vital relationships—technological, spatial, ecological, administrative, and other—as they converge to suggest new opportunities in agriculture today. Recognizing and acting on such opportunities, the text argues, calls on architectural practices of coordinating diverse sets of information and potential stakeholders in order to leverage new social and material formations around the landscapes of food production.

VLOs (Very Large Organizations)

Food production in this context is not merely about agriculture, but inherently about big agriculture. It is a subject of very large organizations: technological, industrial, legal, logistical, economic constructs. Ours is a time to give special attention to the architecture and technological development of very large organizations (VLOs). VLOs are a phenomenon of our day, as the built environments of work, public assembly, agriculture, incarceration, trade, travel, education, and even death join global financial and communications networks. The planning and infrastructure for these built environments demand design, capital, and an order of population magnitude that all must accommodate volatile shifts with spatial and computational stability. Adaptability is at the crux of dealing with diverse users or publics and unprecedented technical, cultural, social, and ecological challenges; and adaptability is where control can give way to engagement and participation.

VLOs are sites for a reconsideration of architectural research practice and they are also action arenas that engage methods of interaction design. This is because they present integrations of architecture and interaction design at unprecedented scales. VLOs are evolving materializations of the administrative, economic, and social orders of our time. They also ask us to revisit notions raised by Garrett Hardin (1968) in the "Tragedy of the Commons," and to consider the social, technological, and ecological changes that have emerged with the recent evolution of very large organizations.

In Hardin's seminal text, the tragedy appears as a result of actions motivated by individual self-interest around the shared resource of the commons, an allegory of sorts that he used to clarify emerging crises caused by the world's population growth. The commons could support many shepherds, but eventually reaches a point of diminishing returns as each shepherd—motivated by individual self-interest—continually brings more sheep onto the pasture. As Hardin (1968, 1248) introduces it: "The population problem has no technical solution; it requires a fundamental extension in morality." The allegory of the pasture represents the larger stress of population growth on the planet as a whole—a phenomenon that has so dramatically intensified since Hardin's own day. But in his essay Hardin also tackled the thorny question of collective organizations and responsibilities around problems at a global scale. If the backdrop to Hardin's commons were stakeholders from government, the sciences, medicine, industry and more, then we can look to a similar mix under the purview of VLOs like big agriculture today.

Agriculture in the United States grew dramatically with the free distribution to farmers of new fertilizing agents and pesticides that were byproducts of the petroleum industry's developments for the military industrial complex resulting from World War II.[2] It took fully thirty years or more for the general public to begin getting comprehensive information about the side effects of these products on water tables, carcinogenic impacts on farmers, and other public health risks. Today we are witnessing the ongoing fallout of big agriculture's reckoning with these discoveries by a broader public; this is a VLO in flux.

A Case Study: Hyperculture—Earth as Interface

Two vitally important fields of work in architecture and computing—in digital fabrication methods[3] and in the development of interfaces between digital and analog systems—can find new forms in their combination with one another. Where computing in architecture was previously most often discussed in terms of computer-aided design *or* interactive systems, today we see more and more work that mixes these interests and areas of research. "Hyperculture: Earth as Interface," an ongoing experiment in the production of landscape rather than building, not only suggests a number of implications for architectural work, but also for the ecological, economic, and urban structures that underlie the project's visible formal and aesthetic orders. Since 2009 I have been developing this in collaboration Virginia San Fratello, an architect. The project is based in California and New York State but targets circumstances that characterize most of the continental United States today. "Hyperculture: Earth as Interface" studies the potential outcomes of modifying a commonly employed information infrastructure for the optimization of agricultural production throughout most of America's heartland, and that same infrastructure's latent flexibility to operate in both "read" and "write" modes, as a means for collaborative input and diversified, shared output.

The Hyperculture project is sited within several contexts—industrial, geographically local, ecological—and within the digital protocols of landscape processing known as "precision agriculture." Today these parameters of precision agriculture typically work together toward the surprising result of repetition without variegation, known commonly as monoculture. After decades of monoculture's proliferation, its numerous inefficiencies have come under broad recent scrutiny, leading to diverse thinking on ways to redress seemingly conflicting demands such as industry's reliance on mass production and automation; the demand for variety or customization in consumer markets; and even regulatory inquiries into the ecological and zoning harms brought by undiversified land use. Monoculture, in short, is proving unsustainable from economic, environmental, and even aesthetic and zoning standpoints. But agriculture's handling in digital interfaces, remote sensing and algorithmically directed fabrication is not.

Repetition and Variation

A first step away from monoculture is the move to polyculture; and then onward to the broader permaculture (a contraction of both "permanent agriculture" and "permanent culture"). Hyperculture goes further, as it captures the unity of agriculture and the information interface. Satellite information networks today drive the creation of landscape, a digital fabrication of landscape affording a high degree of variability. But those same satellites guide us to view the landscape as well; to see, interpret, and understand the earth itself as an interface. We seek to operate on this dual role of the earth as interface.

In "precision farming," GPS now guides tractors to plow and seed fields, instrumentalizing the earth remotely and resulting in a monoculture of cornfields over thousands of acres throughout Iowa and into several other midwestern states. This industrialized agriculture is at the expense of so much. To begin with the ecological fallout, consider the vast water usage; the erosion of topsoil; and the introduction of chemical pesticides and fertilizers into the water table and eventually major water bodies such as the Mississippi. It also costs us all nutritionally, as food products and food subsidies go to the placement of corn in everything we consume, even, as Michael Pollan has pointed out, what our cars consume. In his *Omnivore's Dilemma*,[4] Pollan (2006) examines corn production through a visit to the farm owned by George Naylor, in Greene County, Iowa. For our project, we have done the same, albeit virtually. We have chosen and virtually "bought" a plot that was currently for sale in Greene County, and used it to present a snapshot of what could be technologically, economically, and otherwise viable today.

Earth Interface: Writing (and Reading)

Precision farming uses a mix of digital technologies, including global positioning systems (GPS), remote sensing, and geographic information systems (GIS), to locate, view, analyze, and plan the process of crop cultivation for highly diverse topographic, geological, and climatic conditions across a single farm or a whole region. It relies on software and hardware that has grown steadily more available to the consolidated emergence of big agricultural industry in the United States, and which has in turn led to driving down the price of those same technologies to smaller producers. Through the combination of these digital technologies with the older mechanical technologies of industrial farming—principally, the diesel tractor and its single-seed hopper—monoculture is efficiently propagated across the majority of some eleven midwestern states today. This represents a new scale of land use controlled to results that are monotonous; more importantly, these products are mostly inedible and sustained through a combination of legislated subsidy and technological assistance. This assistance, which we might consider to be "augmented agriculture," is where the Hyperculture project makes its first proverbial incision.

The Hyperculture project makes only minimal adaptations to that existing technical infrastructure. To the digital processing, we add parametric modeling tools such as the common Grasshopper Plug-In (figure 11.1), a graphic patch-based environment that enables the modeling of three-dimensional environments based on dynamically changing ranges of inputs. This can read inputs from local sensors attached to microcontrollers run on an Arduino microcontroller, or be fed live data from false-color

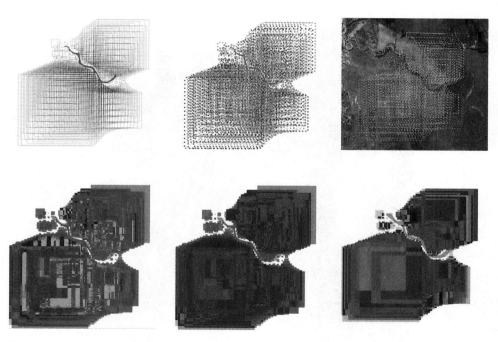

Figure 11.1
Pattern tests address a site featuring high variations in topography, soil sedimentation, moisture content, and nitrogen levels. This pattern is then developed for companion crops and seasonal phasing.

imagery originating with satellite recording (as it is already practiced, see below). This process leads to patterning of diverse plantings, which can in turn be implemented by a simple modification to the tractor hopper to hold multiple seed types (a mechanical modification to this "construction" tool); and an open-source programming platform like Processing to control the multi-seed hopper's actions. This simple combination of moves represent a number of contributions that are native to design, as a convergence of augmented agriculture and contemporary methods of algorithmic architectural design (Meredith, Aranda-Lasch, and Sasaki 2008).

What is gained? These simple modifications to the existing technological infrastructure result in all the desired crop diversity, and to its "legibility" from above, at different scales. The scale of the computer screen can be match-cut to aerial views of landscape; and patterns read for different understandings of use at different scales.

Screen Interface: Reading (and Writing)

Using the same digital tools that produce straight rows of corn, we thus create an overlay of imagery organized by polyculture. These images change appearance at different "zooms" when viewed from above using Google Earth (figure 11.2). This is

Figure 11.2
Hyperculture field result, viewed within Google Earth browser +50 m.

achieved, as we have done, by the creation of .kml and .kmz files, typical map overlay files that can be uploaded, viewed, and searched within the existing browser.

An early test addressed the logocentric potentials of such planning. The question at hand: if crop patterns are to be composed with an awareness of their being viewed or read as a visual artifact, how are they to be composed? In what image is the land to now be created? Logocentrism regards not only the aesthetic dimension of productive landscapes, but also refers to their communicative potentials as well, to their potential iconographic reference to something outside themselves. What is to prevent a Monsanto logo from dictating the material organization of plants, and how is design to take a meaningful position?[5] In producing this first sketch of the Earth Interface in "read mode," we looked to one principle in the classic 1968 short film *Powers of Ten*, by Charles and Ray Eames. As the title suggests, the film is organized by exponential metric zooms. As in our own project, each of these zooms relies on splicing together different types of machine vision: photographic, aerial, satellite, and so on. While the Eames film discovers different visual orders that appear and fade away from one zoom to the next, we chose to design this condition. As a viewer zooms further in, different visual fields and information can be read and understood: these appear first as an image, then as a pattern, then for local differences and finally for local textual and photographic information.

At altitude +2 km, an image of a chimerical plant, a hybrid of its component branches and leaves (figure 11.3) might appear—not a logo, but a figment composed of multiple species: wild garlic, winter rye, vetch, and corn. Zooming in closer to +200 m, each 40-acre parcel breaks into a more abstract pattern of diversely planted rows (figure 11.4). Closer still, the rows are tagged with information and a photo of the crops. This might be thought of as the equivalent of "Street View"; the "Crop View" (figure 11.5). The Google Earth interface and earth interface unite at this moment, as satellites both direct and mediate food production.

This image-based instrumentalization of the Earth's surface has four simultaneous and significant outcomes when implemented: educational; aesthetic; ecological (the reintroduction of plant diversity into an existing monoculture); and economic. The first two are accessed through the browser interface in "read" mode, in which a viewer can study closely and even query local soil conditions, planting cycles, and usage of fertilizers and pesticides to understand the long-range implications of local activities. The latter two are the results of browser input; in the foreseen future interfaces such as Hyperculture will enable visitors to register and participate in the reforming of places near to them, or in which they are personally invested.

The Hyperculture project operates at the intersection of new capacities within existing technologies for information processing, and an open-source platform. Today, more and more users are taking advantage of Google Earth as a resource for examining Earth data previously accessible only to scientists: geology, the environment, conservation,

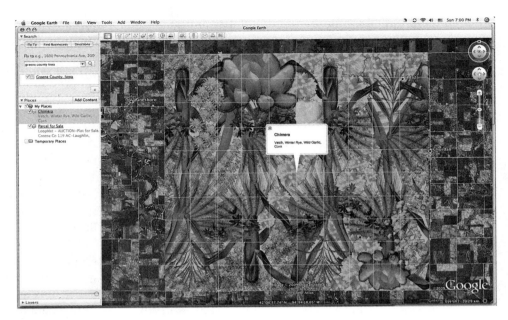

Figure 11.3
Hyperculture chimerical plant test, viewed within Google Earth browser +10 km.

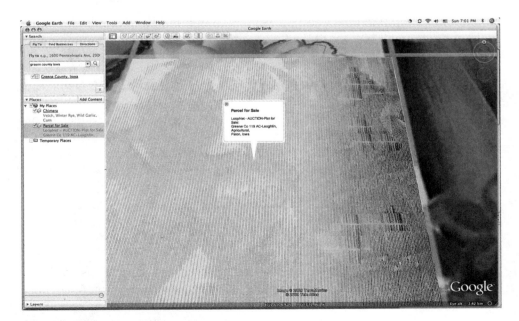

Figure 11.4
Pattern view at altitude +2 km.

Figure 11.5
Pattern view at altitude +200 m.

and renewable energy locations among other things. Such users now also have the option to examine the plant based imagery associated with the Hyperculture initiative in order to learn about the specifics of vernacular planting, soil remediation, agricultural conservation, food miles, and more. The discovery of this showcase on Google Earth will ideally engage users to visit a proposed Hyperculture CSA (community supported agriculture) website, so they may not only be able read the images that are being generated through the polycultural planting, but also to write new landscape as well. Such a participatory or "write" mode has popular precedent in commonplace media today such as the development of music playlists at iTunes and Spotify.

The Hyperculture CSA website will offer more still. Through this platform, consumers, designers, artists, food enthusiasts, gourmets, and gourmands will be able to contribute as members of a CSA model, that is to fabricate the landscape from their computer at home in partnership with a nearby farmer. This would yield new participatory processes of production. Through the CSA model, which is itself enjoying a surge in interest around urban centers in the United States, Europe,[6] and Australia, "grocery shopping" can be thought of seasonally and months in advance, and based on set mixes like playlists or plant lists.

Though satellite imagery can be algorithmically derived, its use for companion crops could be based on diverse inputs like a convergence of atmospheric conditions, an individual farmer's capacities to plant or desires to harvest, collective demand, and shifting nutritional and food-mile data from the United States Department of Agriculture (USDA). The at-home shopper (or artist)—a CSA member—will simply log into the Hyperculture CSA, select a farm in his or her area, and choose the crops to be planted in the field from the "plant list." The farmer will plant the crops most ordered and personally desired, and ultimately deliver them to the at-home shopper approximately six to twelve weeks later. Meanwhile, the CSA member will be able to see through satellite imagery the formation of the plants in the field. In addition to photographing the crops, the satellite imagery is used to analyze the crops and to determine the effects of the planting. For example, to what extent is the planting depleting the nitrogen or phosphorus levels in the soil? Are the crops thriving? If not, the crop density, water saturation levels, soil stresses, and solar conditions can be evaluated through photo manipulation of the ground imagery in order to determine how the next crop can be fine-tuned to respond to each of these specific conditions and fabricate a new landscape. For example, this year, a companion crop planting of corn, pole beans, and radishes may have been planted in the Greene County test plot. Satellite imagery may have indicated that the nitrogen levels in the soil were too high for sustained radish cultivation. Therefore, radishes were removed from the plant list for the next cycle and replaced with a choice of carrots, leeks, or white cabbage. In previous years, the farmer would have speculated and chosen one of these crops to plant himself; this time, however, the Hyperculture CSA allows for mass customization of the fabricated landscape and website users choose to plant 15 percent carrots, 84 percent leeks and 1 percent white cabbage. Their choices can be uploaded to the farmer's processors and all crops planted using a CNC satellite-driven seeder (figure 11.6), which distributes seeds based on percentage allocations of the offsite CSA member's. This process guarantees 100 percent sales for the farmer and shared risk on the part of CSA consumer-members, resulting in higher profits, soil remediation, and maximum crop variation. This is variation in the landscape not only aesthetically but nutritionally as well.

Next steps for this project involve other implications for landscape fabrication and the interface, for which the Hyperculture project is now in a new phase of planning to produce real partnerships and harvest test crops with new modified farming equipment. These will be undertaken with CSA partners on both coasts of the United States and with support from the federal government itself.

Notably, the USDA is responsible for both agricultural and nutritional standards in America and for production and consumption of food. Its website would be a natural location for the public to learn about the relation between things like food miles (planning and production) and nutrition (consumption). Very interestingly, the USDA also collects and distributes crop data layer information (CDLs) as part of its mission, all of which is based on the same remote sensing and machine vision that we implement

Figure 11.6
Hyperculture interface flow diagram.

in Hyperculture.[7] At this time, the USDA is issuing annual grant subsidies for uses of new technologies to support the proliferation of small and independently operated farms throughout the country and to serve smaller, at-risk communities. In short, the infrastructures of the agricultural VLO—technological, economic, governmental, and cultural—are aligning to support a project such as this. At the same moment, individuals and not-for-profit foundations across the country are producing projects that support such a convergence. Examples include the Sarapis Foundation, which aims to distribute "free / libre / opensource (FLO) technologies . . . [as] a human right." Sarapis supports food collaboratives, news services, and the distribution of tools and techniques for food activism. Marcin Jakubowski's Open Source Ecology project,[8] meanwhile, started as a repository for brilliant new designs of CNC-producable, open-source farm equipment of Jakubowski's invention. Demonstrating how accessible and inexpensive the manufacture of food productions tools can be, the website has now become "a network of farmers, engineers, and supporters that has been imagining and creating the Global Village Construction Set, an open source, low-cost, high performance technological platform." Real Time Farms[9] is more about consumption than production: a self-described "crowd-sourced online food guide" nevertheless shares some of Hyperculture's program to let local communities have direct learning from and discourse with nearby food producers.

Within the scope of the Hyperculture project itself, exurban landscape planning drives technical methods around the site planning of buildings but also fosters our understandings of the legal, technical, and biological links between the built environment and public health. The digital interface of Hyperculture can serve as a model for site planning and input on contested sites and public projects within cities and through the processes of building and zoning approvals. This is achieved thanks to the feedback loop of reading and writing, using available technologies to handle diverse demands on a site of public interest. In the context of industrialized agriculture, this work not only negotiates seemingly contradictory demands with diametrically opposed ecological and social outcomes; but also shows the fabrication of landscape as suggestive of other, more architectural applications in the built environment.

Aerospace and Alimentary Architectures

The Hyperculture project builds on the existing technical infrastructure of precision farming by adding parametric modeling and adaptation of the seed hopper mechanisms to the existing reliance on GPS, sensing, and GIS. It is noteworthy that precision agriculture itself is a byproduct of previous technological developments in aerospace. The exploration of Earth atmosphere and sensing technology using satellite cameras (figure 11.7) was initially undertaken by NASA. According to NASA's own account today:

A new generation of farmers is using aerial and satellite remote sensing imagery to help them more efficiently manage their croplands. By measuring precisely the way their fields reflect and emit energy at visible and infrared wavelengths, precision farmers can monitor a wide range of variables that affect their crops—such as soil moisture, surface temperature, photosynthetic activity, and weed or pest infestations. . . .

A number of scientific studies over the last 25 years have shown that measurements in visible, near-infrared, thermal infrared, and microwave wavelengths of light can indicate when crops are under stress (Moran 2000). Using satellite- and aircraft-based remote sensors to precisely measure the wavelengths of radiant energy that are absorbed and reflected from the land surface, scientists can diagnose a wide range of growing conditions. For instance, these data can tell farmers where their crop is thriving and how efficiently the plants are photosynthesizing.[10]

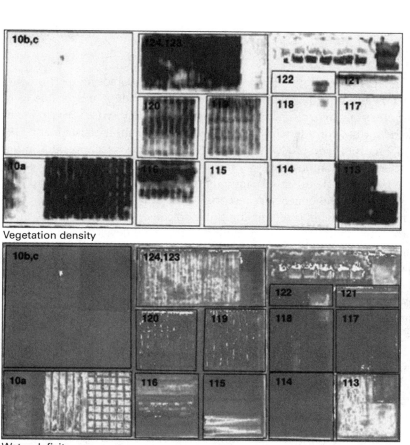

Vegetation density

Water deficit

Crop stress

Figure 11.7

False-color images demonstrate some of the applications of remote sensing in precision farming. (Color can be viewed at the NASA Earth Observatory website.) Source: NASA Earth Observatory. http://earthobservatory.nasa.gov/Features/PrecisionFarming/precision_farming2.php.

Aerospace emerges here for its numerous crucial roles in promoting food production. Just as petroleum development and its government subsidy represent one technological determinant for agricultural practices after the war, aerospace's development during the Cold War has contributed satellite photography, remote sensing, and GPS as three essential tools of food production today. Some of today's technological innovations move to the ground itself,[11] but the "read/write" mode of satellites ties together practices of the farm and the urban consumer in communication, learning, and new protocols for shared risk and benefits in an entirely new way.

Hyperculture is also in support of ongoing efforts to conserve materials in the management of productive landscape and thus found in technical development but independent of legal instruments such as those found in the 2008 farm bill's controversial continuation of subsidies for corn and soy production.[12] NASA and the farm bill share not only a concern for satellite communications and technologies, but also a reliance on federal government for their functioning. A project such as this one therefore needs to account for diverse players and to coordinate their interests as a new sort of architectural work with multiple clients.

The project therefore expands disciplinary boundaries for architecture even as it attempts to unite available technologies and interdisciplinary study of large-scale digital fabrication. It harvests lessons from methods of landscape analysis and design where they have not been recognized as design because aesthetic and experiential factors have not been a consideration since resultant spaces were originally not to be inhabited or even seen. The results discussed above are numerous and might be considered an "alimentary architecture," one that conceives sustenance and sustainability at many dimensions and with new organizational, formal, and material processes and sensibilities—and even with origins in programs for aerospace.

Post-Production

This last note on formal and material processes and sensibilities might be taken lightly, even read as frivolous in a discussion that otherwise rests on the gravitas of food justice, the farm bill, and the imposing powers of big agriculture. Yet today's forms of research, like forms of design practice described at the beginning of this essay, are also deeply hybrid: they mix methods, invite unexpected collaborators, and effect forms of entrepreneurship that are also activism. Moreover, they erase any false sense of boundary that may have previously existed between things like infrastructural work and aesthetics. Today, production and consumption become murky terms in a world of amateur experts and open-source creation. To borrow an observation from the field of art curating: "Shareware does not have an author but a proper name. The musical practice of sampling has also contributed to destroying the figure of the Author, in a practical way that goes beyond theoretical de-construction (the famous "death of the author" according to Barthes and Foucault)" (Bourriaud 2005).

In an earlier time author Leo Marx looked to American novels (Marx 1964) to understand relations between technology and the farm. In cultural production, in other words, he identified our commonly held beliefs and values around agricultural production. Not long before that, the great architectural historian and critic Siegfried Giedion recounted the combined processes of different technologies, and the human impacts, around twentieth-century farming technologies (Giedion 1948). These are two cases of texts that asked where our understandings of relations between design, food, and urban life have come from in the United States. As we consider reading and writing today—not only our own reading and writing but that produced with algorithmic processes and the augmentation of satellite positioning—freedom of information translates to food justice and seeing is consuming is producing.

With newer methods, new responsibilities, and new, more fruitful opportunities than before, design's purview evolves and gives rise to alimentary architectures—as not merely agricultural, but cultural.

GREEN BEANS WITH GINGER, CORN, AND MISO

Makes 4 servings

This is a favorite recipe from my favorite cookbook, the *Three Bowl Cookbook* by David Scott and Tom Pappas of the Yokoji Zen Mountain Center in Southern California. The dish includes a healthy mix of colors and textures, including raw corn. It is remarkably simple to make, and rich in flavors and healthy vitamins—a fitting companion to thinking about a project like Hyperculture!

Ingredients

1 tbsp. peanut or vegetable oil

1 cup fresh raw corn kernels, removed from cob

1-inch piece of fresh ginger, peeled and minced

4 cups green beans, parboiled and chopped

½ cup sake or dry sherry

2 tbsp. white miso

Preparation

1. Heat the oil in a skillet or a cast-iron pan. Add the corn kernels, minced ginger, and green beans.

2. Sauté until the corn and beans are heated through.

3. Add the sake and continue to cook until the liquid is reduced by half.

4. Once reduced, stir in the miso.

5. Remove from the heat and serve.

Notes

1. The journal *Computers and Electronics in Agriculture* is published in Amsterdam by Elsevier Science.

2. See for instance trade documentation like the collected papers in *American Chemical Society: Agricultural Application of Petroleum Products* Volume 7, June 17, 1952.

3. This has been subject of many publications and conferences over the past ten years in particular, with perhaps the annual ACADIA (Association for Computer Aided Design in Architecture) conference being the most prominent convergence of voices.

4. While not as widely known outside the United States, Pollan's book has contributed within the country inestimably to the popularization of food discourse. Coming from the field of journalism, Pollan has succeeded in making complex stories about food accessible to new and broad publics in the past five to ten years, to the point that his books are now sold in airport bookstores everywhere.

5. Consider for instance the provocative 1995 "Parkhouse Carstadt" project by NL Architects, in which rows were arranged to create a giant Mazda logo to be seen from planes landing at a nearby airport. Drivers would not pay but rather be paid to park, as they incrementally contributed to the design's completion.

6. See chapter 1 in this volume, by Joon Sang Baek, Anna Meroni, and Giulia Simeone.

7. http://www.nass.usda.gov/research/Cropland/SARS1a.htm.

8. http://opensourceecology.org.

9. http://www.realtimefarms.com.

10. http://earthobservatory.nasa.gov/Features/PrecisionFarming.

11. A somewhat messianic account of several methods can be found in this fascinating review by Thomas Frey. http://www.futuristspeaker.com/2009/11/the-coolest-profession-on-earth-next-generation-agriculture.

12. http://frwebgate.access.gpo.gov/cgi-bin/getdoc.cgi?dbname=110_cong_bills&docid=f:h2419 enr.txt.pdf.

References

Bourriaud, Nicolas. 2005. How to inhabit global culture (aesthetics after MP3). In *Postproduction: Culture as screenplay: How art reprograms the world*. 2nd ed. New York: Lukas & Sternberg.

Giedion, Siegfried. 1948. Agriculture in full mechanization. In *Mechanization takes command, a contribution to anonymous history*. New York: Oxford University Press.

Greenfield, Adam, and Mark Shepard. 2007. *Urban computing and its discontents*. New York: The Architectural League of New York.

Hardin, Garrett. 1968. The tragedy of the commons. *Science* (3859) (13 December): 1243–1248.

Marx, Leo. 1964. *The machine in the garden: Technology and the pastoral ideal in America*. New York: Oxford University Press.

Meredith, Michael, Aranda-Lasch, and Mutsuro Sasaki, eds. 2008. *From control to design: Parametric/algorithmic architecture*. Barcelona: Actar.

Pollan, Michael. 2006. *The omnivore's dilemma: A natural history of four meals*. New York: Penguin Press.

Waldheim, Charles. 2006. *The landscape urbanism reader*. New York: Princeton Architectural Press.

12 The Allure of Provenance: Tracing Food through User-Generated Production Information

Ann Light

For the last three days we have been snaking among forested backwoods, talking to coffee planters and hearing tales of shaky futures, falling returns and an aging workforce whose children have left for the city—the gambles of growing a volatile crop.

As a measure of protection, diversification is rampant. The hills support coffee only if grown in the shade and the need for cover suggests supplementary crops. Cardamom, pepper, vanilla and oranges accompany the main harvest. Pigs, cows, and poultry provide manure. Honey also helps the economy while promoting the pollination of the coffee blooms. Several generations of planters around Koorg have built an ecology largely in tune with existing patterns of nature. Launched on the back of a requirement to use at least one dressing of organic products every three years is a new breed of farmer now managing whole estates with sufficient environmental vigilance to have organic certification for crops. But most of the growers here are in charge of 10 hectares or less and fluctuations in coffee prices, an invasion of the white stem borer insect, bank foreclosures and a powerless position in the trade chain have contributed to a rise in suicide.

Is the shade-grown coffee of Karnataka a commodity? It is traded as one, but elsewhere the quality and conditions of production would make it a speciality product, sold at a premium to discerning importers and European aficionados. In fact, Signor Illy of the eponymous Italian firm has his own room of bean sorters at the local grading plant to capture the best of the Karnataka harvests for his company. Bigger producers attract roasters to visit directly and so secure a reasonable margin over rival coffees from Vietnam or Columbia. Meanwhile, the smaller planters gather together in federations and join the marketing cooperative.

These growers talk to us over excellent coffee, in their plantations, their living rooms and offices. The stories are similar: we need to make the quality of our coffee known; we are growers that live in harmony with our environment; we may one day qualify as a carbon sink; the wildlife that populates our plantations makes this a special area of diversity; we take a paternal interest in our workers' wellbeing; the last few years have made us question our enduring commitment. Because, without the trees that punctuate the fields and the coffee that needs them, the growers could put in cash crops like rubber. They speak with a mixture of resignation and temptation.

(Visit to meet coffee producers, Ann Light, Fair Tracing project, 2007)

The narrative above is one of many stories that could be told about the provenance of shade-grown coffee. I include it here to share what we experienced in going to research a provenance tool in the raw, an account of the project team's first trip to meet the growers of southern India in 2007. But it also combines typical features of marketing: it is both evocative and emotive; it personalizes and yet conveys general facts about the product; it defies readers to think of coffee as a commodity by giving a context to its production. It is a dramatic tale of coffee's source. Significantly, it was not written by anyone involved in the production, retailing, marketing or reviewing of coffee, but by a spectator, and yet it is publicly available, posted on a blog. It might be the first account of Indian coffee you see. As such, it is part of the networked world, just one of a patchwork of voices contributing to the definition of the product.

The Internet is renowned for assisting in the distribution of accounts from widely flung authors to a worldwide audience, with an increasing elaboration of websites to host, display, and field content, and the appearance of social media—a range of structures around which people can organize, create, and make new relations (see, for instance, Gauntlett 2011). In theory, anyone in the value chain—from producer to consumer—can post their own data about profit distribution, product review, or production story. In practice, things are never so simple. The Fair Tracing project asked about the new opportunities and barriers to telling these stories.

Provenance as Marketing

Producers and food outlets such as supermarkets and restaurants have long been in the business of selling nourishment as part of a narrative in which consumers are active agents, not only in the visceral process of eating and drinking, but in the meaning-making practices that accompany it.

Already in 1959, Sidney Levy suggested that "sellers of goods are engaged, whether willingly or not, in selling *symbols*, as well as practical merchandise" (1959, 117). Now marketing theorists increasingly understand retail in terms of co-constructed experiences, activating the resources that different marketplaces provide to create meaning. Bernard Cova and Danielle Dalli (2010, 477) identify four experiential stages (Cova and Dalli 2010, 477):

1. Pre-consumption: planning, day-dreaming, imagining the experience.
2. Purchasing: the sales environment, choosing the item, payment, packaging.
3. Core consumption: sensation, satiety, (dis)satisfaction.
4. Remembered consumption: when photos, movies, etcetera are used to relive past experience.

In the fourth stage, media are presented as a mainstay of evoking and building on the experience, involving consumers in a productive capacity that goes beyond reflection and interpretation to making and sharing.

Throughout, as consumers speculate and reflect, concerns about provenance arise. Peter Oosterveer and David Sonnenfeld (2012, 233) note that: "more and more consumers define food quality beyond simply the objective characteristics of food products and include (in)direct environmental, animal welfare and social impacts as well. Environment-related food concerns have supplemented and adapted consumer worries about the safety, quality, taste and price of food." In other words, provenance information—about where something has come from and how producing it has impacted the environment—has grown in significance in how people regard choices of purchase and how they feel about themselves in having made one choice over another.

Inevitably, this has influenced the positioning of food products over the last few years. The sales pitch is now more likely to include some provenance information to engage with these concerns and suggest the consumer can play crusader. Peter Jackson, Polly Russell, and Neil Ward (2011) describe how, as part of a company's marketing program, a fictional association is developed between a range of chicken products and the name of a small market town that speaks to "countryside imagery and nice places" to suggest the poultry was considerately reared. Further, each label has the name of the actual farmer stamped on it, and also an endorsement from a *typical* farmer with picture—indicating 100 percent traceability and showing the company's pride in these rearing methods.

The project discussed here took the example of two drinks, coffee and wine. It considered the narratives that can be built around particular instances of production. What distinguished the research reported here is (1) a focus on driving ethical consumption and production, and (2) experimentation with the new elements that the Internet can bring to tracing provenance for marketing purposes. The dimensions along which the proposed tool could be ethical and the trade-offs that result in trying to manage social and environmental production issues are described elsewhere (Light 2011). Here, we consider how the Internet can be used to link up the chain between producer and consumer in an information journey that accompanies the production journey and redefines provenance.

Fair Tracing: Overview of a Provenance Project

There are many provenance management systems in use in the world, but most are private, employed to guarantee food accountability. These processes make it possible to isolate a batch of food and take it out of circulation if necessary. At their most advanced, tracking systems are based on networked auto-identity capture, using technology such as radio frequency identification (RFID) tags, and allow goods to be managed without human intervention—in terms of accountability, but also in terms of logistics: following goods between suppliers along the supply chain; monitoring storage and sales; and handling reordering. Small businesses trading with global or highly regulated players, as European supermarkets are, must have their own systems for managing traceability. At the level of the micro-enterprise, few use automated means. Even digital spreadsheets are rare; batches are recorded in ledgers by hand.

Fair Tracing (www.fairtracing.org) was a three-year UK-led project to research a tool to bridge the divide between Global North consumers and Global South producers by using tracing technology to enhance ethical trade (Light and Anderson 2009). The research was intended to promote smaller producers, especially those in developing countries, and help them make a presence in global markets by using the Internet to manage and share information, and to include consumers (and potential consumers) in a visible provenance management process. Oosterveer and Sonnenfeld (2012) observe that globalization is having a major impact on food production and distribution and a world perspective is becoming increasingly necessary. Although there has been a recent rise in "buying local," the last century has seen longer chains of distribution and many consumers become disconnected from the sources of their food. The prospective tool might therefore increase and facilitate choice for both ethically minded traders and consumers who wish to understand and discuss the origin of their purchases.

Our research question was: Would small-scale producers in developing countries be able to use an open source Fair Tracing tool to better understand the value chains they operate within and distinguish their product by adding production information and communicating directly with consumers? If so, what would consumers make of it?

There were several parts to the research; the next paragraphs give a sense of what we did, while following sections discuss the provenance design aspects in greater detail.

Production

We observed two small enterprises, chosen for their diversity as case studies: the shade-grown coffee growers of Karnataka, India, featured here; and a Chilean Fairtrade wine collective. We mapped the passage of products from growth to point of sale and looked at the transitions they went through (Light 2010). We discussed and observed use of information communication technology (ICT), current marketing practices, and information flows. We used co-design workshops to elicit business opinion on new approaches (Light, Kleine, and Vivent 2010).

Among a wealth of data, we learned the high number of trading hands Indian coffee goes through before consumer sale; the fact that beans are roasted and crushed, making identifiers difficult to attach; and the limited quantity of (still paper-based) information that travels with it.

Consumption

We ran surveys of British Fairtrade /green / general consumer attitudes to coffee and wine and ethical food issues. We organized researcher-accompanied shopping trips to farmers' markets to discuss provenance (Light, Wakeman et al. 2010) and to supermarkets to explore the range of coffee and wine on offer and discuss selection processes and what information might be useful at point of purchase. We interviewed British and Indian (i.e., domestic) coffee drinkers about their social and political concerns over a prototype interface (Light 2011).

We learned that people have different priorities depending on the product. For instance, the stated impact of price, distance traveled, and health effects differ between wine and coffee. We learned that people want different information at different times in the buying and consumption process (in the shop, in preparation, at the table) and most people do not want detailed data, but a way of choosing between available products. We also found that shopping behavior is not consistent with people's claims about it. For instance, shoppers at the farmers' market took ethical provenance on trust because they liked the "production narrative" told by the salesperson, whereas they said they needed specific information for more commercial enterprises (Light, Wakeman et al. 2010).

System/Product Design

It is possible to organize a tracking system using the assignment of numbers and chronology to each actor and event (from coffee growing to drinking); attach each actor and event to a particular item (i.e., an instance of a product); and then attach related information, such as stories, to that item, with evidence of where the story has come from. But how far would this process be of use in situ along the value chain and for interested third parties? We evaluated the suitability of technologies such as RFID tags and barcodes for collecting data from micro-business back ends. We looked at the intervention of other players in the flow of products and information and considered means of including them. We ran trials with a barcode reader on a phone for scanning production data and displaying it to consumers.

We learned that our partners are ill-prepared to use even barcodes to tag their crops as they collect so little information in usable form, but that there is some consumer appetite for a swipe mechanism that displays data. And although intermediary / third party organizations, such as the Fairtrade Labeling Organisation or Oxfam, are difficult to include in a formal model of the product journey, such organizations are sufficiently interested in the concept to consider participating to overcome these technical shortcomings.

Interaction Design

We worked with an international syndicate to review ethical markers and scoring systems for products, looking for a standard, clear way of representing information about provenance and ethical production processes. We developed interface designs based on social networks, maps, and timelines and compared the narrative aspects which each stressed (Light 2011). We considered the politics of representing different parts of the world to each other through a single interface and what challenges this raises in linguistic and cultural terms.

Some of these discussions inform the next sections.

Researching Provenance Content

The project looked at how operations data transmitted along the production chain could be enhanced by images, audio and/or text, to provide information for potential purchasers at point of sale and elsewhere. Material might include stories of corporate social responsibility, community or environmental impact, details of the economic and environmental costs of creation, the individual creator, their working environment and pay, through the steps of its transport and processing, to reviews by consumers. The creation of audiovisual and narrative material would be in the hands of the actors along the chain with no central editorial control determining what might be published or by whom.

Chains of production with environmental, economic, and/or social strengths would benefit from this kind of exposure. Our partners were groups of small growers with an ethical production story to tell, struggling to survive in an international context of big distributors, as the chapter opening relates. So there were compelling stories from the start of the chain.

Other features were not exactly production stories, but affected the fate of the crop. Because of tax laws, the smallest coffee growers benefit from selling the coffee beans they produce to middlemen as soon as they are grown. If they take them to the curing works (the next stage in production, where beans are graded—see figure 12.1), they are liable to pay 25 percent tax on them. Bigger estates producing quality coffee pay the tax because they can keep their beans distinct during grading, brand them as high quality "single estate" beans, and recoup the outlay. The smallest producers, delivering tiny batches of coffee, cannot maintain the identity of their beans, however good the quality. Their crop—and, with it, detailed provenance—is lost into a pile of similar beans. Stories like this show the value of a provenance tool to the producers with smallholdings, who might be able to leverage identity information for better profits. They are also tales in themselves.

Thus, coffee involves many collaborators in getting to market. At each stage profit must be made. There are stories to tell about how the chain is connected as well as individual production stages along it. These could include stories of the traders functioning as middlemen, who aggregate goods but add no value. They could include growing, grading, roasting and grinding coffee. We found that these details were all of interest to consumers, especially when related to a breakdown of profits.

Telling these stories has a politics closely related to that of the chain: "because strong dependencies exist, producers feel constrained to tell a politically conservative story

Figure 12.1
Many coffee beans arrive at the curing works and lose their provenance.

for fear of offending distributors and those further up the chain, making it awkward to share facts that would otherwise be in the producers' best commercial interests to make public" (Light 2010, 34). Power relations along the chain are all-pervasive. For instance, even to collaborate with our wine producers, we had to secure permission from their importers in the United Kingdom, who had, in turn, to clear participation with their distributors. Each provenance story might need distributor sign-off, a daunting prospect in the busy world of the micro-producer.

What became evident in hearing stories of the chain is that defining provenance in this context is not merely a matter of identifying origins in an absolute sense. It is not the provenance note of an antique, dating it and assigning it an original maker and/or owner. Provenance emerges as more than the sum of the stories, but as something constructed by the actors along the production chain in the associated chain of information. The way that actors present themselves and their actions (or leave gaps) as part of any publically visible sequence becomes a de facto history of the product. And even though the result may be a semi-arbitrary aggregation of data, anecdotes, myths, and interpretations, it starts to affect the sociotechnical systems around it as well as attitudes to the thing itself. This distinguishes it from the marketing described above, where the farmer of the chicken becomes the remote icon for the product in a controlled branding exercise that regards provenance as origin, rather than patchwork journey.

With these insights in mind, we next reflect on a detail of presenting provenance and how media aesthetics came to play a part in the research.

Experiencing Provenance

In this section, we juxtapose producer and consumer media values to reveal a mismatch in expectations, not only of information, but of presentation styles.

At a trade conference in Bangalore in 2007, we watched the local coffee industry discuss their latest coffee marketing initiative. A short film had been made featuring a Coffee Swamy (figure 12.2), an animated coffee deity shaped like a coffee bean (http://www.biganimation.com/showcase2.html), which narrated the excellence of the local crop and production methods against shots of the beautiful landscape where coffee is grown. To our eyes, it was kitsch; the swami appeared twee to us, and the narration clichéd, making the film inelegant despite the beauty of the cinematography and the environment. But the producers loved it. Clearly there were different aesthetic values at play. But perhaps there were issues of identity at stake. Showing a well-produced video of modern cultivation processes in beautiful landscapes placed the producers in a glamorous role with a glossy marketing story to tell. This was traditional marketing at its zenith.

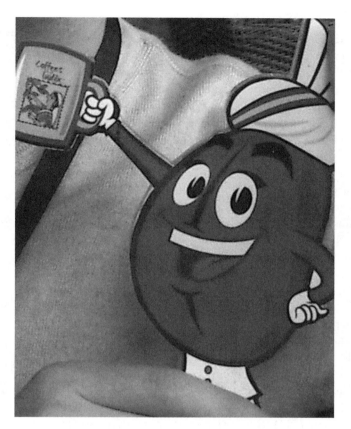

Figure 12.2
The Coffee Swamy

Toward the end of the project, we were back in Bangalore, interviewing coffee drinkers about their expectations of the Fair Tracing interface. The coffee drinkers of Bangalore are a sophisticated group: middle class, discerning about their purchases, and knowledgeable about coffee because many of them have relatives in the business. The people in the group we interviewed were, furthermore, selected for their capacity to use computers and speak to us in nuanced English, which is common in middle-class India, but reflects an education that is not available to all. So we were interviewing a particular sociocultural echelon.

British and Indian consumers were asked to look at a map, a timeline, and a social network and say what each suggested to them by way of representing the "story of coffee from bean to cup" (see Light 2011). Our discussions extended beyond the immediate interface details to wider aspects of presenting the value chain. The method of interviewing we used did not initiate conversation—the interviewees chose which interface to view first from a still image of each and explained what they saw to us, working through all of the interfaces in this way until they felt ready to make a comparison and a more general assessment of the tool and its potential.

As we talked about showing ethical production with one interviewee, the topic of video came up. The interviewee had strong views. She explained what she would expect to see: "Not like a documentary, just little snapshots. I'd rather like to see what happens in the fields; the auctions. I think we are used to real videos now with YouTube. We would look carefully and if they've not made it carefully we would be able to spot 'Hey look, there's a kid picking beans there.'"

Not only would this allow her to see for herself into the production practices claiming to be ethical, it would also offer credibility and charm: "If it's unprofessional, we know it's in real time. Actually, sometimes unprofessional is more attractive than professional."

Here we recognize an appreciation of current media aesthetics. She likens the material she expects to that of the "real videos" of YouTube posted by people recording their own material as it happens and then sharing it. She is averse to anything that might be staged. And she can see the value of this aesthetic for presenting the material we are discussing: it would allow consumers to make their own judgments; it would reflect the authentic situation of a small producer with limited media resources ("unprofessional," made "in real time"); and thus it has the potential to be more attractive. In contrast to the values of the producers at the conference (and also of those we interviewed—see Light 2011), this Indian consumer showed media literacy skills and taste that resembled the UK consumers we met.

As mentioned, the project team had earlier witnessed the coffee growers' marketing film with the thought that high-end production values and cute storylines would not satisfy the British market with an appetite for provenance, a view supported by findings from both the farmers' market and supermarket. Here was evidence that the film would not satisfy parts of the domestic market for coffee either.

The growers believed that anything but highly professional quality could be misunderstood. The trendy roughness of a viral campaign showing/imitating homemade content could be viewed as reflecting on the quality of the coffee, suggesting "third world" quality emanating from India. Ignoring this dynamic, our interviewee embraced the contemporary global aesthetic of DIY media. This tells us something of the complex social stratification of Indian society, but more importantly here it also tells us more about the "look and feel" of provenance. The aesthetics of authenticity currently involve a grainy and overtly amateurish quality. This is distinct from the photos of a named *typical* farmer on chicken products, which go part way (or indeed, the named *typical* coffee grower glossily featured on the wall of the coffee shops selling Fairtrade coffee in the United Kingdom).

This contrast raises further issues about the role of the chain actors as contributors and in whose name they publish. The postings of most social networks speak individual to individual and the "amateurish" aesthetic is a leveler, but here the wider business context comes into play. The tool might provide freedom to act, but the system around it provides constraints, from the politics of the message to that of the styling.

Discussion

The Fair Tracing tool was conceived as a means of leveling the playing field between small producers and big companies who are able to commission tracking technologies and exploit product data (usually without sharing it). Fair Tracing was about openness—using nonproprietary technologies, putting no limits on participation and making information available to all. Information itself would be marked with its provenance, making data and stories publically available to producers, consumers and other actors, with the opportunity to trace them.

At the outset, it was hoped that by allowing those at the bottom of the pyramid to tell their own stories and present information about how things are made, the tool would give a platform for ethical information and potentially drive more ethical production and purchasing practices. As research developed, we identified many obstacles to producing this information, from time constraints to politics to language, media, and digital literacy issues (Light 2010). All could be surmounted, but at a cost.

Nonetheless, by encouraging content generated by a multiplicity of users, the tool offers the opportunity for people to join in making and sharing information. Although there is dispute as to whether knowledge drives change in practice (see the summary in Oosterveer and Sonnenfeld 2012), there is more evidence that participation does drive change. As we move away from the tightly controlled editorial model of classic marketing and use a tool that assembles diverse voices, we surrender end points, honed messages, and coherence for a semi-arbitrary assemblage of contrasting aesthetics and messages which, at its best, enacts the product journey and offers a dynamic version of provenance, full of trade-offs, gaps, and new developments. This is provenance we can share in. The description at the start of this chapter, then, is my story, not that of the growers; it is the record of *my* distress. This is a long way beyond the acknowledgment of co-constructed meaning upon which marketing hinges, beyond perhaps Cova and Dalli's (2010) fourth stage of consumption experience, of media constructed as a form of remembering. It is a highly active and reflective role aligned with the quizzing sensibility of the Indian interviewee who wants to be able to *see for herself*. It holds the beginning of new relations with our food and the producers of it.

Bernard Cova, Robert Kozinets and Avi Shankar (2007) suggest that, as consumers, people live in a specific social and historical situation that places them in a co-dependent relationship with commercial culture, giving them the material with which to create social and cultural identities, and leading them to absorb and resist the prepackaged meanings of the marketers. That social and historical situation is, of course, open to change. Our investigations show, once more, that significant cultural differences exist in this situation across nations and between them and these fuel different interpretations of the same phenomenon. Since the interpretation, once incorporated in the chain, takes on a weight of its own, these too become actors in defining provenance and in revealing our potential roles to us.

We are in a phase of intense experimentation. Marketing is traditionally dominated by the interests of the retailer and producer with the goal of selling more goods. Every cultural trend has been exploited to this end and there is little doubt it will happen again. But there is a new rhetoric; caught by design company IDEO:

The political economists argued two centuries ago that economies are based on three functions: production, consumption, and exchange. Since that time, first production, then consumption has taken its turn as the driving forces of value creation. As the consumption era draws to a close, exchange is emerging as the next driver of economic growth. (Mamut and Powell 2009)

The Fair Tracing project, in disclosing more about the constitution of provenance as contemporary technology can deliver it, begins a conversation about how knowledge of provenance exchanged along the chain might change the means of production and the chain itself, and the role of consumers as well as producers in determining it. With the future of food uncertain and concern about production issues growing, provenance has never been a more important story. Perhaps the principal contribution of the Fair Tracing project then has been to ask some difficult questions about creating and presenting the new dynamics of provenance possible, drawing on the ubiquitous threads of information of the Information Age.

Acknowledgments

The author thanks the UK's Engineering and Physical Sciences Research Council for funding and acknowledges the work of Fair Tracing's Ian Brown, Ashima Chopra, Dorothea Kleine, Apurba Kundu, Helen Le Voi, Cote Montero, Macarena Vivent, and Christian Wallenta.

MULLED WINE

Makes 1–2 quarts

This drink is traditionally served in winter in Britain. It dates back many centuries to a time when wine, oranges, and spices were luxury items. While most British people now have access to all these ingredients throughout the year, only sugar is regularly produced in the United Kingdom. It is interesting then to consider how to make the drink while applying social and environmental principles. Immediately, the familiar trade-offs are highlighted. Most products can be bought as Fairtrade, but this means longer journeys to the United Kingdom, since Europe is outside the remit of Fairtrade, and oranges, for instance, could be easily sourced from Spain. And which crops have been grown by ecologically friendly means? Most products can be bought as organic, but many cannot be found as Fairtrade and organic at the same time.

Ingredients

1–2 quarts red wine

1–2 oranges, depending on size

Freshly grated nutmeg

A mix of whole spices, which might include cloves, cinnamon sticks, ginger, star anise and cardamom

1–2 cups sugar

(Measures do not need to be precise)

Preparation

1. Pour wine into a large saucepan on low heat and add nutmeg.
2. Place whole spices in a muslin bag and knot, or put straight into the pan.
3. Shave off several strips of orange peel (avoiding pith) using a peeler or knife; then juice the orange.
4. Add juice of orange and strips of peel to the pan.
5. Add sugar and stir till dissolved.
6. Keep mixture simmering below boiling point for 5 minutes and then strain and serve in large wine glasses. Add thin slices of orange for decoration.

(The hotter the mixture is kept, the more alcohol will burn off. So, for a strong brew, turn the temperature down after the wine reaches the just-below-the-boil point.)

References

Cova, Bernard, and Daniele Dalli. 2010. The linking value in experiential marketing: acknowledging the role of working consumers. In *Handbook of marketing theory*, ed. Pauline Maclaran, Michael Saren, Barbara Stern, and Mark Tadajewski, 476–493. Thousand Oaks, CA: Sage.

Cova, Bernard, Robert Kozinets, and Avi Shankar. 2007. *Consumer tribes*. Oxford: Butterworth Heinemann.

Gauntlett, David. 2011. *Making is connecting: The social meaning of creativity, from DIY and knitting to YouTube and Web 2.0*. Cambridge, UK: Polity Press.

Jackson, Peter, Polly Russell, and Neil Ward. 2011. Brands in the making: A life history approach. In *Brands and branding geographies*, ed. Andy Pike, 59–74. Gloucester, UK: Edward Elgar.

Levy, Sidney J. 1959. Symbols for sale. *Harvard Business Review* 37 (4):117–124.

Light, Ann. 2010. Bridging global divides with tracking and tracing technology. *IEEE Pervasive* 9 (2):28–36.

Light, Ann. 2011. The politics of representing cultures in ubiquitous media—Challenging national cultural norms by studying a map with Indian and British users. *Personal and Ubiquitous Computing* 15 (6):585–596.

Light, Ann, and Theresa Anderson. 2009. Research project as boundary object: Negotiating the conceptual design of a tool for international development. ECSCW 2009: *Proceedings of the 11th European conference on computer-supported cooperative work*, ed. Ina Wagner, Luigi Ciolfi, Hilda Tellioglu, Ellen Balka, 21–42. London: Springer.

Light, Ann, Dorothea Kleine, and Macarena Vivent. 2010. Performing Charlotte: A tool to bridge cultures in participatory design. *International Journal of Sociotechnology and Knowledge Development* 2 (1):36–58.

Light, Ann, Ian Wakeman, John Robinson, Anirban Basu, and Dan Chalmers. 2010. Chutney and relish: Designing to augment the experience of shopping at a farmers' market. [New York, NY: ACM.] *Proceedings of OzCHI* 10:208–215.

Mamut, Tatyana, and Ashlea Powell. 2009. The exchange economy, IDEO: http://patterns.ideo.com/issue/the_exchange_economy

Oosterveer, Peter, and David A. Sonnenfeld. 2012. *Food, globalization and sustainability*. London, UK: Earthscan.

13 Beyond Gardening: A New Approach to HCI and Urban Agriculture

Tad Hirsch

A growing number of designers and technologists have recently become interested in urban agriculture. Several projects seize on the promise of aeroponics and hydroponics to enhance food production in confined urban spaces. Windowfarms,[1] for example, is a window-mounted hydroponic kit that allows city dwellers to grow herbs and produce in cramped apartments, while Dickson Despommier's Vertical Farm Project[2] proposes high-rise buildings dedicated to agriculture to be erected in the heart of major cities. Other projects bring automation and remote sensing capabilities to small-scale, urban farming operations. Botanicalls,[3] a student project from New York University's ITP program was an early entrant that alerted people by phone when their plants were in need of water. More recently, Luke Iseman has developed and begun marketing Garduino,[4] a control system for small-scale cultivation projects based on the open-source Arduino electronics prototyping platform. Still other efforts bring the easy networking capabilities afforded by social media platforms to connect urban producers and consumers. Fallen Fruit[5] and Forage Oakland[6] promote urban gleaning by producing maps of fruit trees and other edibles on public lands. Websites like Yardsharing[7] and Hyperlocavore[8] connect homeowners willing to share their gardens with would-be growers in need of land.

Despite the spate of recent activity, the human-computer interaction (HCI) community has largely overlooked urban agriculture as a potential site for research and development. While there have been a few studies, they have generally been small in scope and have lacked theoretical richness. As a result, they have so far failed to inspire widespread interest by HCI researchers.

In this chapter I revisit the topic of urban agriculture and its potential for HCI. Drawing on urban sociology, I provide a new, expanded definition of urban agriculture and offer a framework for thinking about the different design opportunities it presents. I also provide a case study of Sunroot Gardens, an urban agriculture project that was active in Portland, Oregon, between 2006 and 2010. I consider Sunroot Gardens as an ambitious attempt to reimagine an entire food supply from a decidedly urban perspective, and describe the range of strategies, tactics, and specific design interventions it entailed. Finally, I reflect on the Sunroot Gardens initiative to make several observations about design, technology, and urban agriculture more generally.

My goals in presenting this work are generative. I hope to support designers and engineers currently engaged in urban agriculture projects, and encourage new entrants to the field. At the same time, I highlight the social and cultural dimensions of urban agriculture as a challenge to the ways that HCI typically thinks about cities, technology, and the environment. In short, I hope to articulate a set of theoretical concerns that can deepen our thinking and expand our practice.

Understanding Urban Agriculture

Despite growing international interest, urban agriculture has been largely overlooked by the HCI community. While several papers have explored relationships between small-scale food production and interactive systems (e.g., Hirsch et al. 2010), the few studies that have focused on urban agriculture have been small in scope and have lacked conceptual depth. As a result, they have so far failed to inspire widespread interest by the HCI community.

Prior work has generally understood urban agriculture to mean something akin to "food production within city boundaries," (Odom 2010), and have drawn a rough equivalence with gardening. HCI researchers have examined community gardens (Goodman 2009), and have proposed technical interventions to promote water conservation among home gardeners (Pearce and Murphy 2010). However, urban agriculture itself remains largely under-theorized, which perhaps reflects a lack of theoretical work underpinning urban informatics more generally (Williams, Robles, and Dourish 2009).

The emphasis on community and home gardeners has a kind of intuitive appeal. However, it reveals misconceptions about both cities and agriculture that, ultimately, limit HCI's potential contribution to the problem of urban food provisioning. In the first instance, it privileges urban form over urban experience, describing cities in geographic, rather than social, terms. At the same time, it reduces food provisioning to cultivation, focusing on the growing of vegetables but not the distribution and consumption of foodstuffs. Taken together, these two misconceptions offer a very narrow vision of urban agriculture that so far has failed to spark significant interest among HCI researchers.

From Urban Form to Urban Experience

It is common within HCI and urban informatics to conceptualize cities as discrete, geographically bounded entities that are easily discernable from, say, rural and suburban locales. This way of conceiving urban space has been largely rejected by recent urban scholarship (Williams, Robles, and Dourish 2009). As Henri Lefebvre (1996) observed, the urban/rural divide is rendered incoherent in an advanced capitalist society, where the economic and social lives of cities, town, and countryside are inextricably linked. A more contemporary reading of urban space is Manuel Castells' (2000) widely employed description of cities as nodes in complex, global networks through which people, materials, capital, and information "flow."

Nowhere is the blurred boundary between rural and urban space more apparent than in food production. In an age where over half the world's population lives in cities, and where food distribution networks span the globe, farmers around the world are largely producing foodstuff for urban markets. At the same time, the growth of cities around the world has had a pronounced effect on the landscape, changing land-use practices and transforming historically rural areas into mixed-use, peri-urban zones. Accordingly, crop selection, planting schedules, organic certification programs and a host of other farming practices are shaped by the demands and tastes of urbanites. In a very real sense, all agriculture is urban.

The emphasis on geographic boundaries is based on a broader and similarly misplaced emphasis on the material aspects of the built environment. This is not to say that the distinctive physical characteristics of urban space don't present unique challenges for urban agriculture. Constrained growing spaces, contaminated soils, and poorly understood microclimates certainly affect farming practices. However, reducing "the city" to these sorts of material concerns misses the forest for the trees (or perhaps, the city for the buildings). As Lewis Mumford (1937) noted, the city is far more than a "purely physical fact." Cities are not simply a collection of buildings and roads; rather, the physical infrastructure is the material substrate through which the city as "social drama" is enacted. For Mumford, the city is "a theater for social action." This sentiment is a foundational principle for urban sociology, and is echoed in Lewis Wirth's (1964) description of a distinct urban culture that arises from, but which cannot be reduced to, essential urban conditions of scale, density, and social heterogeneity. Rather, urban sociology teaches us to shift our focus from urban form to urban culture and experience.

The notion of a distinctive urban culture is contentious; there have been numerous and largely unsuccessful attempts in the history of urban sociology to define it. Castells (1983) argues that there is not a single, distinctive urban culture that characterizes life in all cities at all times, but rather that cities contain and indeed are shaped by competition between various and often contradictory cultures. At the same time, we can acknowledge Wirth's (1964) observation that the unique material conditions of cities—scale, density, and heterogeneity—exert an influence on the structure of social relations. Wirth describes this as a de-privileging of community in favor of association. Wirth's argument is based on Emile Durkheim's distinction between mechanical and organic solidarity; the former describes social relationships based on identity, often expressed through shared belief, occupation, or demography. Organic solidarity, in contrast, is predicated on interdependence: actors may have little in common with each other and indeed may hold values in direct opposition to one another, but they share interests (in the sociological sense) and are engaged in complementary tasks that compel them to form ties and coordinate their activities (Durkheim 1997). Durkheim argued that organic solidarity characterizes social relations in complex societies, which are exemplified by the industrial city.

Importantly, the contingency of urban social relations is often an explicit topic of concern among urban residents, and is therefore a constitutive component of urban experience. Decisions about with whom we associate, and how, are often construed as matters of choice and expressions of values and are therefore rife with ideological commitments, personal preferences, and the like. It is thus a common feature of urban experience for such mundane activities as shopping and transportation to figure prominently in public (and sometimes contentious) social and political discourses.

From Cultivation to Food Provisioning

In the same way that cities are more than a collection of physical assets, so too is agriculture more than an amalgamation of soil and plants. Prior work has focused on gardening by individuals, often as a recreational activity. This is true even of the community garden studies, which examined gardening projects in which individuals managed their own subplots. While there was a fair amount of socializing and resource sharing among participants, these projects nonetheless support community gardening as a parallel rather than collective activity.

However, agriculture isn't simply the practice of growing and harvesting crops (i.e., cultivation). Nor is it typically considered to be primarily recreational in orientation. Rather, agriculture is concerned with supplying edible, healthy food to the public (i.e., provisioning). Agriculture includes selecting cultivars, preparing soil, growing and harvesting crops, and preparing and marketing the resulting products. It is concerned with entire supply chains including the cultivation, processing, and distribution of food. Agriculture is thus also a fundamentally social activity, bound up with labor, economics, institutions, and culture in ways that extend well beyond an individual and her garden.

The point is not that cultivation doesn't matter. Clearly, it is a crucial component of agricultural practice. Nor do I intend to dismiss gardening as a form of agriculture. It is. However, focusing solely on cultivation and gardening misses a much larger opportunity for HCI practitioners to help reinvent urban food provisioning. As I demonstrate in this chapter, there are opportunities for design intervention throughout the food supply chain.

An (Expanded) Definition

Bringing together the preceding sections, we begin to see urban agriculture as a social enterprise that deals with urban form and culture, and exists at the intersection of food and urban systems. We might define urban agriculture as the practice of producing and distributing food within cities, in ways that reflect the distinctive character of urban life. It is distinctive "in location, in economic motive, in type of product, in the use and distribution of harvests, in actors involved and in the types of technologies used" (van Veenhuizen 2003).

As an expression of urban experience, urban agriculture is a collective project undertaken by groups of actors bound by interdependence. Because these relationships are to some extent voluntary associations, they facilitate a reflection on and articulate of the common values and concerns upon which they are based. They encourage a kind of intellectualism. It is therefore unsurprising that urban agriculture has long been associated with social movements. This is particularly true of the United States and Europe, where the growth of urban agriculture in recent years has largely occurred in the absence of severe food shortages or other strong material drivers.

A Framework

Having defined urban agriculture, let us now consider several areas of design inquiry within it and begin to articulate specific design issues and HCI opportunities within each.

As we have already established, urban agriculture includes various practices of cultivation, which are primarily concerned with the growing and harvesting of crops and animal products. Design issues here tend to center on the material challenges of urban food production, including microclimate variation, soil remediation, irrigation, pest control, crop selection and planning, and the like. For the HCI practitioner, much of the design opportunity in this space involves measurement, control, automation, and decision-making supports. As Elizabeth Goodman (2009) and Will Odom (2010) have both observed, there may be some resistance to technical interventions at this scale; however, it is also worth noting that in both, cases studies were carried out with hobbyist communities for whom the meditative and physical aspects of gardening were of paramount concern. Their findings may not generalize to commercial or other non-hobbyist practitioners.

Urban agriculture is also concerned with food distribution, which primarily involves delivering the products of cultivation to consumers. Design issues here include marketing, transportation, storage, business models, and so forth. For HCI, the design opportunities here include information services to assist publicity and customer engagement, transportation planning and monitoring, consumer demand modeling, price setting, and transaction support.

Urban agriculture also requires organizations that coordinate resources and activity, as well as institutions and policies that govern urban food supply systems. Design issues include organization design, human resources, policy development, and compliance. HCI practitioners will find opportunities in simulation and modeling, workforce management applications, and compliance monitoring technologies.

Finally, urban agriculture depends on culture—social conventions that shape patterns of food production and consumption. Design here focuses on creating and reinforcing values that support urban agricultural programs. HCI practitioners are encouraged to develop arts and education applications that engage the public and promote such values as sustainability, localism, and the livability of cities.

In considering this framework, there are two points worth emphasizing. First, the various modes described here are deeply interrelated. It is both possible, and in many cases, desirable for designers to work at several scales simultaneously. Second, the design space is not strictly a technical one. While I have attempted to make this essay more accessible to the HCI community by calling out opportunities for technical interventions, it should be made clear that the design challenges presented by urban agriculture call out for sociotechnical innovation that coordinates activities by machines, people, and institutions.

To better see how these theoretical concerns play out in practice, let us now turn to a specific example of a recent urban agriculture initiative.

Case Study: Sunroot Gardens

The following case study was developed through a mixed methodology. I conducted participant observation between 2008 and 2010, when I was a paid subscriber and occasional volunteer with Sunroot Gardens. During this period, I conducted semistructured and informal interviews with staff and volunteers, and analyzed newsletters, websites, and journalist accounts. Additional semistructured interviews were conducted by telephone and email in 2011.

Project Description

Sunroot Gardens was an urban agriculture project that was active in Portland, Oregon, from 2006 to 2010. It attempted to engineer an entire food supply chain, situating production, distribution, and consumption within a single neighborhood. Conceptually, Sunroot brought together several strands in the budding urban agriculture movement, including repurposing decorative yards for garden space (de-lawning), sharing space in private gardens with neighbors (yardsharing), human-powered transportation, and community-supported agriculture (CSA).

The project was initiated by a former tech worker and media activist named Kollibri terre Sonnenblume, or simply "Farmer K," in response to the apparent precariousness of the industrial food system. Sunroot Gardens was born in the midst of dramatic increases in food prices that led to shortages and famine around the world, and spawned riots in over a dozen different countries. Sonnenblume (in Brooks 2009) has described the Sunroot Gardens workers as "urban survival farmers." As Farmer K saw it, the industrial food system was on the verge of imminent collapse, and Sunroot Gardens was intended to provide a reliable source of food in the coming post-industrial era.

Sunroot's primary orientation was toward cultivation and the practicalities of growing vegetables. The organization included several well-versed amateur horticulturists who routinely referred to plants by both their Latin and common names, and could expound in great detail on the history, culinary, and medicinal properties of hundreds of fruits, vegetables, and herbs. The group members devoted many hours to study, seeking out and reading obscure and out-of-print texts on traditional agriculture and the Pacific Northwest. They were also keen observers who took careful notes on planting schedules, environmental conditions, and yields to better understand the unique characteristics of each of their gardens.

Despite the obvious passion and expertise in growing vegetables, though, actual gardening represented only a portion of Sunroot Gardens' offering. Sunroot was in various parts a distributed farm, a CSA, an innovation lab, and an ongoing social experiment. In short, it attempted to address the entire system of food production, distribution, and consumption through a decidedly urban lens.

As a distributed farm, Sunroot Gardens spread across several dozen front yards, backyards, empty lots, and sidewalk strips in Portland's Southeast neighborhood. At its peak, Sunroot Gardens managed approximately forty-five distinct gardens, which were tended by a mix of five paid staff and roughly twenty-five volunteers. Sunroot Gardens grew a wide variety of produce, including vegetables, herbs, and medicinal plants. Sunroot also cultivated a small tobacco crop. Transportation—including deliveries of tools, produce, and workers—was almost exclusively by bicycle. Produce was distributed to approximately thirty CSA subscribers, as well as shared by volunteers and sold at a neighborhood farmers' market. Sunroot Gardens was one of the few CSAs in the Portland area to operate year round.

As an innovation lab, the project served as an incubator or launch pad for several projects. In 2008, Sunroot Gardens members initiated the Staple Crops Project to cultivate wheat, quinoa, and other grains. As at Sunroot Gardens itself, most of the labor was done by volunteers using hand methods. During its first year, Staple Crops harvested six hundred pounds of wheat. In 2009, Sunroot Gardens partnered with several other Portland urban agriculture organizations to open the Hawthorne Farmers' Market, described as "Portland's only market by and for ultra-local produce and urban farmers" (Hawthorne Urban Farmers' Markets n.d.). The Hawthorne Farmers' Market provided a commercial outlet for urban farmers who had previously distributed their produce exclusively through CSAs. Sunroot Gardens also initiated the SE Portland Seed Bank project, with the object of creating "varieties of crops that are well-suited to our soils, climate and farming methods in adequate quantities to meet the demand of local food security" (SE Portland Seed Bank Project n.d.). This seed bank was an outgrowth of Sunroot Gardens' prodigious seed-saving and exchange program, which over the years had amassed a collection of seeds for hundreds of varieties of vegetables, grains, herbs, and medicinals. These were gleaned from a variety of sources including Sunroot Gardens' own gardens where each year, a portion of the crop was reserved for seed production to provide a basis for subsequent seasons. Several of the varieties represented in Sunroot Gardens' seed library had been grown in the same location for multiple generations and would likely represent unique strains. Sunroot Gardens typically hosted several seed-exchange parties each year, where attendees were encouraged to take seeds in exchange for all manner of goods and services.

Finally, as a social motivator, Sunroot Gardens embodied an ideal. Vegetables are commodity products: one carrot is more or less exchangeable for another. What made Sunroot Gardens carrots special was the fact that they were grown locally, delivered by bicycle, and, in lieu of a plastic bag, came wrapped in paper printed with a set of narratives about sustainability, autonomy, and alternative culture. It was the sum total of these stories and associations, combined with actual, edible produce, that constituted Sunroot Gardens' offerings, and it was precisely the intangible aspects of the operation—and the way that they reinforced such cultural values as sustainability and localism—that made Sunroot Gardens valuable to its constituents.

A Principled Approach

Sunroot Gardens exemplified the expanded definition of urban agriculture presented above. While it engaged in cultivation, its aim was to reinvent entire systems of food provision. It was also a thoroughly urban project that relied explicitly on notions of solidarity and interdependence to motivate participation and coordinate action.

The Sunroot Gardens case also reveals a tight connection between values and design. The project began with a strong and explicitly articulated sense of purpose: to provide food to a small number of people in a secure and sustainable manner. It's worth dwelling for a moment on exactly how the organization thought about security and sustainability. As is probably apparent, Sunroot Gardens' approach to sustainability was both holistic and austere. The organization approached sustainability as a social, economic, and environmental concern. Sunroot Gardens was intended to be (or rather, to become) a closed-loop system that could maintain itself without consuming additional resources. The organization took a similarly hard-line approach to food security. In addition to providing affordable food year-round and mitigating risk of disruption by extreme weather and natural disaster events, Sunroot Gardens sought to limit dependency on institutions that the organization considered unreliable, including both the industrial food and the global financial systems. These goals were obviously ambitious, and should be thought of as aspirational; they may or may not have been fully achievable, but the project was shaped throughout by an honest attempt to get as close as possible to them.

In order to fulfill the mission it laid out for itself, Sunroot Gardens engaged in design activity across the framework described above. Sunroot Gardens' members developed a set of material and information artifacts that supported cultivation and distribution. These were many and varied. They designed gardens, deciding which crops to locate at each site, and planned elaborate layouts of vegetable beds, herb gardens, walkways, and tool storage at each location. The Sunroot Gardens team devised a number of low-impact, human-powered tools and techniques for growing, processing, and trans- porting produce and equipment including simple machines for threshing wheat, homemade drip irrigation systems, and sturdy bicycle trailers. They designed spread- sheets to plan and monitor planting, maintenance, and harvesting schedules. The group also developed communication systems involving email lists, websites, and printed newsletters to share information including harvest reports, pickup and work schedules, recipes, and personal musings with subscribers, volunteers, and the com- munity at large.

They also developed several organizations and their attendant policies and proto- cols. Some of the protocols were encoded as formal rules for governing payment and distribution schedules, for instance. However, many protocols remained informal. Part of the work of creating Sunroot Gardens included developing a culture that encour- aged, among other activities, recipe sharing and gift giving among participants.

Sunroot Gardens' design activity was shaped by a set of five principles that flowed from the organization's mission and its approach to sustainability:

1. Reduce or eliminate fossil fuel. Sunroot Gardens believed, as do many in the sus- tainability movement, that one of the key problems with conventional agriculture is its reliance on fossil fuels for fertilizer, transportation, and powering machinery. Sunroot Gardens adopted a strategy of eliminating or reducing fossil fuel use in all aspects of its operations, across the entire supply chain. This strategy was manifest in a variety of tactics including the reliance on bicycles for nearly all transportation and general preference for hand tools and human labor over power machines. Similarly, commercial fertilizers, which depend on petroleum-intensive production processes, were not used; soil was instead remediated through compost and crop selection.

2. Build resilient networks. Farmer K was a tireless networker, building a network of formal and informal relationships between individuals and organizations that pro- vided a more or less stable social structure that operated Sunroot Gardens and its various spin-off projects. This network also enabled resource and skill sharing, and encouraged knowledge transfer; as Farmer K put it, "There aren't any trade secrets."

The network-centric approach extended to Sunroot Gardens' approach to land use, which emphasized multiple small lots rather than a single, larger farm. In all cases, networks provided the organization with resiliency. A volunteer might drop out of the project, or a homeowner might remove her garden from cultivation. In either case, the organization could continue with relatively little disruption. It is worth noting that, although Sunroot Gardens officially ceased operations when Farmer K relocated in 2010, it inspired and directly assisted the creation of several other urban agriculture projects in Portland that assumed responsibility for many of the plots that Sunroot originally cultivated. The Hawthorne Urban Farmers' Market also remains in operation.

3. Simplify supply chains. One of the key factors contributing to food insecurity is the reliance on complex supply chains that are vulnerable to disruption (for example, by extreme weather events). Complex supply chains also impede traceability, making it difficult to contain contamination and food-borne illness outbreaks. As with farmers' markets and other CSAs, Sunroot Gardens sought to reduce supply chain complexity, bringing producers and consumers closer together physically and socially. This strategy was intended to improve food security, reduce the operation's overall carbon footprint (by reducing the distance that food had to travel), and foster social bonds between farmers and clients. In many cases, the distinction between producer and consumer was utterly obliterated, as much of the produce was distributed among volunteers and paid laborers; conversely, many subscribers also helped with planting, harvesting, and other tasks.

4. Reduce or eliminate the role of money. Spurred by concerns about financial insecurity among Sunroot Gardens workers and supporters, and by a conviction that the global financial system is on the verge of collapse, Sunroot Gardens sought to separate itself from the formal economy as much as possible. To some extent, this was achieved by severely limiting expenses. Farmer K himself led a very modest existence, and Sunroot Gardens relied heavily on donated equipment and volunteer labor. Also, a percentage of the annual harvest was allowed to go to seed, providing the biological foundations for next year's crop without requiring cash outlays for seed. Measures to promote financial independence went further than reducing costs, though. Sunroot operated largely on a barter economy, which greatly reduced the role of money in the organization. Instead, produce itself served as a kind of currency, and was traded for labor, garden space, and various goods and services.

5. Promote biodiversity. The industrial food system relies on a relatively small number of varietals that have been optimized for such factors as uniformity of color, long shelf life, and transportability. While these strains enable great efficiencies in the food system, monocultures are vulnerable to disease and predation. As a result, their cultivation typically requires significant use of pesticides, which have been linked with groundwater contamination and adverse health impacts. Sunroot Gardens developed

a strategy of crop diversity as a bulwark against the vulnerabilities associated with monoculture agriculture. Crop selection privileged heirloom and local/regional varietals, which reduced the need for pesticides and limited the need for fertilizers and irrigation (the culinary appeal of this crop selection did not go unappreciated, either). Pest control was also achieved through crop rotation, handpicking, and cultivating "beneficial" insects (like spiders) that prey on unwanted insects. Sunroot Gardens also maintained a seed bank and encouraged seed exchange with other growers, which supported greater biodiversity throughout the city.

In sum, Sunroot Gardens attempted to reimagine urban food provisioning in a principled, systematic manner. Participants engaged in deliberate and explicit reflection on their goals and ideological commitments. This directly informed practice at several scales, beginning with the development of an organization and its governing protocols and extending through a set of strategies, practices, and ultimately, physical and informational artifacts. Ultimately, project participants devised a unique and innovative approach to the cultivation and distribution of produce within a city, and developed an organization that embodied, reinforced, and promoted their values.

Connecting Values, Design, and Local Context

Sunroot Gardens situated design activity within a broader effort to reimagine an entire food system. It is worth observing that many of the strategies, tactics, and artifacts outlined above are common in the urban and sustainable agriculture movements. What makes Sunroot Gardens noteworthy is that it had a broad and comprehensive vision for urban-centric food provisioning that extended well beyond simply cultivating herbs and vegetables. As such, it helps us understand where the producing of foodstuffs fits within a supply chain, and illustrates one possible model of how a distinctly urban food system might be constituted. Sunroot Gardens helps us shift our focus from the materiality of cultivation to take in the whole of the food system. This complicates the design space, situating efforts to, say, optimize yields in constrained environments within a broader design effort. Design inquiry thus expands beyond questions such as "how much can be grown in a quarter-acre urban plot" to "what's the most environmentally, culturally, and economically sustainable way to provide foodstuffs to contemporary urban residents?"

The response to the sustainability question is, of course, largely shaped by local context. Sunroot Gardens was very much a product of its environment. The reliance on bicycles, the refined culinary affectations, and the strong counter-cultural identity, for example, all reflect long-established aspects of Portland's Southeast neighborhood culture that might not make sense in other parts of the world. And this is as it should be: urban agriculture is a highly localized practice that should look different in Portland than it does in, say, Havana, where, for example, an equally ambitious urban agriculture program was undertaken in the 1990s. Here, "sustainability" was understood in terms of the end of the Cold War and decreasing support from the Soviet Union to bolster the Cuban economy. Accordingly, urban agriculture was promoted by the central government, and was bound up with notions of national identity (Moskow 1999).

I don't intend to argue the merits of either Sunroot Gardens or the Cuban solution. I simply observe that although these initiatives were both based on notions of sustainability, the ways in which sustainability was conceived and therefore acted upon was highly dependent on local context. The point, for designers, is to recognize that the way we understand sustainability both shapes our practice and is highly contingent. It therefore behooves us to make our conception of sustainability an explicit part of our design practice.

This brings us to a broader observation about goals and values. Sunroot Gardens was rife with ideological commitments. It wasn't simply a technician's attempt to optimize food distribution systems or to wrest greater efficiencies out of small-scale agriculture; it embodied a set of values and political ideals. It was as much a product of and participant in several social movements (sustainability, local foods, anti-corporate) as it was an attempt to do farming. Indeed, the project was initiated by and embodied a radical critique of capitalist food production. The influence of social movements and ideological commitment was particularly evident in the way that notions of solidarity and autonomy undergird many of the design decisions that were made. Consider the primacy of trust and individual autonomy in shaping social relations within the Sunroot Gardens organization. For example, food distributions were handled on the honor system. On collection days, when produce from across the land network was brought to a central location for pick up, subscribers were encouraged to take as much as they desired from Sunroot Gardens' bountiful offerings but requested to not take more than they needed. Pick-ups were often unsupervised; subscribers were also given a map of the various garden plots with descriptions of what had been planted at each, and encouraged to harvest their own food in between distribution days (at least at plots whose owners didn't mind subscribers showing up unannounced to root through their gardens). As described above, there was also a privileging of barter and labor exchange over cash transactions. Sunroot maintained a special collection day for volunteer workers and other "friends of the farm." On several occasions, particularly following the 2008 global financial collapse, Farmer K expressed an interest in phasing out paid subscriptions altogether and discussed ideas for an all-volunteer/barter system.

The observation that design activity embodies a set of ideological values is hardly new (see, for example, Winner 1989). The point here is not only that design is political, and not only that designers have a moral obligation to examine their assumptions (Friedman 1996). I would also argue that having a well-thought-out and explicitly articulated understanding of one's broader goals deepens design work, and expands practice. The connection between theory and practice is highlighted in the Sunroot Gardens case, where we see a direct link between deep consideration of social and environmental issues, and specific design decisions about tools, organizations, and policy. While the specific designs produced by Sunroot Gardens may by highly localized, the practice of connecting critical thinking with design action is widely applicable.

The foregrounding of political concerns within the Sunroot Gardens experiment is also distinctly urban. This is not to say that rural or suburban agriculture doesn't have a politics. However, the centrality of ideology in the Sunroot Gardens project, and the way that these politics were expressed—especially over such spatial concerns as land use and transportation—was quite consistent with the tenor of other urban social movements (Castells 1983) and with other contemporary urban agriculture projects, many of which are bound up with concerns about sustainability, local culture, and anti-corporatization. It is also consistent with the tendency discussed above for urban culture to explicitly link everyday activities with broader social concerns and commitments. We see these commitments expressed by contemporary urban agriculture movements in myriad ways, including preferences for low-impact technologies, organizational models based on trust and solidarity, and counter cultural symbolism (often expressed as a sort of neo-hippie aesthetic). To be sure, there is no particular reason why urban agriculture has to be associated with progressive political causes. However, it would seem that being associated with some political cause—particularly one with a strong spatial component—is part of what makes urban agriculture urban.

Conclusions

I have suggested that designers reconsider urban agriculture as not simply an attempt to grow plants in small spaces, but rather as an ambitious program to reinvent food provisioning from a distinctly urban perspective. I have offered a framework for thinking about various scales of design intervention, including cultivation, distribution, organizational design, and cultural engagement. I have also suggested that designers also need to engage in a critical and explicit articulation of the goals and principles underpinning their efforts, as situated in local context. Finally, I have offered Sunroot Gardens as an instructive, albeit low-tech, example of what urban agriculture might look like under this expansive definition. Reflecting on the full scope of Sunroot Gardens' ambition and activities helps us to think critically about other, more focused efforts, and to think creatively and expansively about new ways in which to bring information and communications technologies to bear on the problem of urban food systems.

Looking at Sunroot Gardens, we see intervention points for design across the supply chain. We also see opportunities for technology artifacts to reduce dependencies on fossil fuels, build resilient networks, simplify supply chains, mitigate the influence of money, and promote biodiversity. The broader perspective offered by Sunroot Gardens also highlights possibilities for new, combinatory artifacts that bring together different aspects of the food production system. For example, we might imagine connecting remote sensing with social networking technologies to produce garden monitoring systems that not only track when crops are in need of tending, but also interface with a network of volunteers to coordinate the necessary labor and equipment—or, perhaps, explore the use of CSA planning and management tools that enable networks of producers to coordinate crop selection and timing, and offer variable subscription packages that allow consumers greater flexibility and choice.

Importantly, Sunroot Gardens was a profoundly urban project, and as such, can help us more fully appreciate the social and cultural dimensions of the urban agriculture movement. In particular, we see how urban agriculture is bound up with ideological commitments, particularly with regard to spatial issues. Further, we see how an approach to agriculture as a collective project undertaken by groups of individuals bound together through solidarity relationships based on interdependence leads to an emphasis on organization and social form within design activity.

It is my hope that broadening our conception of urban agriculture and its possibilities for design will help HCI practitioners to envision new ways of bringing their expertise to bear on the pressing problem of urban food supply. Ultimately, I think, the lesson that Sunroot Gardens has to offer technologists interested in urban agriculture is a simple one. It is both an invitation for new practice and a cautionary tale that, as we develop our gadgets and collect our data we ought not to lose sight of the bigger picture. Ultimately, urban agriculture isn't about wresting ever-greater efficiencies out of tiny growing spaces; nor is it about data collection and analysis. The goal is nothing less than to invent new ways for cities to feed themselves, sustainably and equitably. Our job, as technology designers, is to provide the material and organizational support that will enable this vision to take root and grow.

SAUTEED KALE

Makes about 4 servings per pound of kale
Sunroot Gardens was one of the few year-round CSAs in the Portland area. While the available produce varied greatly throughout the year, kale was always on offer, and for good reason: it's tasty, nutritious, and easy to grow all year long, even through the Pacific Northwest's notoriously sun-deprived winters.

This simple recipe is based on a Brazilian preparation, and works equally well with collards, chard, or other leafy greens. In winter months, I often finish it with a splash of red wine, which gives a slightly heartier flavor. In the summer, I simply squeeze a bit of lemon over the greens for a lighter, more acidic taste.

Ingredients

Garlic

Kale

Olive oil

Red wine (optional)

Preparation

1. Rinse kale and remove spines.
2. Lay the leaves out flat, and stack them one on top of the other.
3. Roll up the stacked leaves, so that they resemble a cigar.
4. Slice the "cigar" crosswise into ribbons.
5. Heat olive oil and garlic in a sauté pan.
6. Add the kale and sauté for a few minutes, until tender and bright green.
7. Transfer kale to serving dish.
8. Swirl a splash of wine into the pan to capture any leftover bits of garlic, and pour over the kale.
9. Season with salt and pepper, and serve.

Notes

1. See www.windowfarms.com.

2. See www.verticalfarm.com.

3. See www.botanicalls.com.

4. See http://garduino.dirtnail.com.

5. See http://fallenfruit.org.

6. See http://orageoakland.wordpress.com.

7. See http://yardsharing.org.

8. See http://hyperlocavore.ning.com.

References

Brooks, Jonathan. 2009. Community supported agriculture in the city. *Examiner.com*. http://www.examiner.com/article/community-supported-agriculture-the-city.

Castells, Manuel. 1983. *The city and the grassroots: A cross-cultural theory of urban social movements*. London: Edward Arnold.

Castells, Manuel. 2000. *The rise of the network society*. 2nd ed. Malden, MA: Blackwell Publishing.

Durkheim, Emile. 1997. *The division of labor in society*. New York: Simon and Schuster.

Friedman, Batya. 1996. Value-sensitive design. *Interaction* 3 (6):16–23.

Goodman, Elizabeth. 2009. Three environmental discourses in human-computer interaction. In *CHI EA '09 Proceedings of the 27th International Conference Extended Abstracts on Human Factors in Computing Systems*, 2535–2544. New York: ACM.

Hawthorne Urban Farmers' Market. n.d. About Hawthorne Urban Farmers' Market. http://hawthornemarket.wordpress.com/about.

Hirsch, Tad, Phoebe Sengers, Eli Blevis, Richard Beckwith, and Tapan Parikh. 2010. Making food, producing sustainability. In *CHI EA '10 Proceedings of the 28th of the International Conference Extended Abstracts on Human Factors in Computing Systems*, 3147–3150. New York: ACM.

Lefebvre, Henri. 1996. Town and country. In *Writings on cities*, ed. Eleonore Kofman and Elizabeth Lebas. Malden, MA: Blackwell Publishing.

Mumford, Lewis. 1937. What is a city? *Architectural Record* LXXXII (November):58–62.

Moskow, Angela. 1999. The contribution of urban agriculture to gardeners, their households, and surrounding communities: The case of Havana, Cuba. In *For hunger-proof cities: Sustainable urban food systems*, ed. Mustafa Koc, Rod MacRae, Luc Mougeot, and Jennifer Welsh, 77–83. Ottawa: International Development Research Centre.

Odom, Will. 2010. "Mate, we don't need a chip to tell us the soil's dry": Opportunities for designing interactive systems to support urban food production. In *DIS '10 Proceedings of the 8th ACM Conference on Designing Interactive Systems*, 232–235. New York: ACM.

Pearce, Jon, and John Murphy. 2010. "Living on the hedge": Creating an online smart garden watering community. In *OZCHI '10 Proceedings of the 22nd Conference of the Computer-Human Interaction Special Interest Group of Australia on Computer-Human Interaction*, 420–421. New York: ACM.

SE Portland Seed Bank Project. n.d. The mission. *SE Portland Seed Bank Project*. http://sepdxseedbank.wordpress.com.

van Veenhuizen, René. 2003. Micro-technologies for urban agriculture. *Urban Agriculture Magazine* (10):1–3.

Williams, Amanda, Erica Robles, and Paul Dourish. 2009. Urbane-ing the city: Examining and refining the assumptions behind urban informatics. In *Handbook of research on urban informatics: The practice and promise of the real-time city*, ed. Marcus Foth. Hershey, PA: IGI Global.

Winner, Langon. 1989. Do artifacts have politics? In *The whale and the reactor: A search for limits in an age of high technology*, ed. Langdon Winner, 19–39. Chicago: University of Chicago Press.

Wirth, Louis. 1964. *On cities and social life*. Chicago: University of Chicago Press.

14 Hungry for Data: Metabolic Interaction from Farm to Fork to Phenotype

Marc Tuters and Denisa Kera

While food might at first seem an unconventional topic for human-computer interaction (HCI) design research, our intention is to exploit this apparent mismatch so as to question some of its basic assumptions, in particular around shared notions of the user (or eater) and her local context. To eat and to know what we eat, where it comes from, and how it interacts with our molecules—to be at once a beast and a god—this is the paradox that we try to reconcile through a concept we will refer to as metabolic interaction.

In this chapter we use design theory to explore material, social, and discursive practices around food and consider new models of agency through a discussion of our design research. We offer the metaphor of metabolism as a way of thinking about interfaces that go beyond conventions regarding stabilized definitions of users and contexts in order to trace interactions across a variety of scales. Within food politics the concept of the local is often treated as being axiomatic. This chapter therefore begins by reflecting on the issues of locality by looking at how new mobile and locative forms of HCI make possible a form of politics based on "the trace." Our metaphor of multi-scalar metabolic exchanges thus builds largely on actor-network theory's (ANT) concept of traceability in order to account for how heterogeneous networks are constituted through performance. As applied to HCI, the concept of metabolism refuses both to assume the user as the sole or even primary agent in an interaction and to assume a stable context as the ground for such interaction. Metabolism as a metaphor allows us to design whole systems, which we refer to in terms of collectives, whose constitutions undergo dynamic processes while they interact. We are not only beings that eat, we are also beings that feed, or metabolize, and that will eventually, in turn, be metabolized by some other process ourselves. Rather than users-centered design and context-aware systems, we thus offer the concept of metabolism as a model of how to design for multi-scalar interactions from the genetic to the planetary (from farm to fork to phenotype) in order to trace the topology of food systems and to consider these configurations, in terms that ANT scholar Bruno Latour refers to as "cosmopolitics." To this end, our approach combines Bruno Latour's sociology of the trace

(2005) with Brillat-Savarin's politics of the table (2004) in the service of a form of design practice that Paola Antonelli (2011) refers to as "design for deliberation."

To address this process of interaction across scales, we develop new narratives, in the form of design scenarios and prototypes, within which to make sense of the social and political but also ontological implications of new agencies related to technological and scientific innovation. Thus, while the first part of the chapter is devoted to developing the concept of metabolic traceability in relation to emerging locative technologies, in the second part we present examples from our own practice based on a specific design scenario titled "Diet-Tribe Diatribes." In this section we discuss a series of prototypes we developed in 2010–2011 inspired by current-day extreme food practice and speculative fiction including a social network for sharing leftovers ("FridgeMatch") and a meal in which we treat the eater's individual gene type as the end user ("Eat What You Are: Dining on Biopolitics"); we conclude with a recipe drawn from the latter project that illustrates our concept of metabolic interaction at the micro scale of phenotypic interaction. We begin, however, at the macro scale in order to consider how design interaction can be used to connect our bodies with larger social systems, and, following from that premise, how our notion of metabolic interaction might contribute to the problems in our understanding of core HCI concepts of the user and of context.

But Is It Local?

Consider the following anecdote taken from the recent television program *Portlandia*.[1] A waitress approaches two diners, asking them if they have any questions about the menu:

Diner 1: "I guess I have a question about the chicken, if you could just tell us a little bit more about it."
Waitress: "The chicken is a heritage breed, woodland-raised chicken that's been fed a diet of sheep's milk, soy, and hazelnuts . . ."
Diner 2: ". . . and this is local?"
Waitress: "It is."
Diner 1: ". . . is that USDA organic, or Oregon organic, or Portland Organic?"
Diner 2: ". . . the hazelnuts? Are they local?"
Diner 1: "How big is the area where the chickens are able to roam free?"

The waitress leaves for a moment returning with a dossier on the chicken, which lists personal traits, daily diet, and so forth. It does not satisfy the diners. Rather than eat the bird, they choose to visit the ranch on which it was raised, a back-to-the-land commune, where the politics of their food overtake them, and they end up joining a cult.

This anecdote evokes the increasingly politicized relationship between people and things, and the failure of standard measures to adapt to this complexity. The diners in this scenario are not so much hungry for chicken as they are hungry for data. In fact, the performance seems to be more a meditation on their place in the order of things or in relation to the cosmos than a simple consumer choice. The humor seems to arise in the juxtaposition between their ravenous appetite for data and the inappropriateness of their measurement: is it local? This hunger for data can be connected to a trend that emerged from within media art in the past decade, which sought to augment real-world analog experiences with digital data. What was originally an art practice, referred to as "locative media," has in recent years become a commercial industry in which context-aware devices allow users "to personalize and control their experiences of physical spaces" (Gordon and de Souza e Silva 2011, 11). This new concept of locality is based on the ability, through context-awareness, to associate data with places and with objects so that their logistics may be traceable. Given these innovations, it seems the locavore's obsession with determining the exact provenance of food and every quanta of carbon that goes into its production would be better served by a traceability interface than by this concept of "the local."

Interfaces that connect our fork to the farm support the fantasy of a local and controllable metabolism in an age of globalized exchanges. For Bruno Latour (2005) however, "no place dominates enough to be global and no place is self-contained enough to be local." From this perspective, the local and the global are both essentially fictions. Furthermore, it has even been argued that an unreflexive "construction of the local," in relation to food networks, can paradoxically become bound up in a biopoliticial project of governmentality, in effect "reproduce[ing] the very neoliberal subjectivities that [it] seek to oppose" (Harris 2009, 55). In this critique, the problem occurs when the concept of local scale is equated with "good," as opposed to a recognition of how it is constructed in relation to complex networks, thereby risking the depoliticization via commodification of the activist impulse at the core of locavore food politics. As the aforementioned sketch makes preposterously clear, "the local" is, as Latour (2005, 193) states, always "oozing out" to a variety other places and other scales. To show the dominance of networks over nodes and points, Latour curated a media art show titled "Making Things Public" in 2005, with Peter Weibel at the ZKM Center for Art and Media in Karlsruhe. The show presented itself as an "experimental assembly of assemblies" (Latour and Weibel 2005). One project in particular, simply titled The MILK Project,[2] illustrated how a new experimental apparatus, could give rise to a new social theory. Portraying the provenance of a Dutch cheese from a cow in Latvia, the artists used the concept of GPS traceability and data visualization to give voice to a previously silent entity. The MILK Project went on to win the 2005 Golden Nica for Interactive Art at Ars Electronica, initiating a traceability genre in which artists and designers use HCI technologies to represent objects as assemblies.

In Latour's framing, such traceability techniques reveal that "politics might not be so much about opinion as about things—things made public" (Latour, n.d.). His claims are based on an empirical method that he has pioneered over several decades of science studies in which he traces the often unstable webs of associations at the core of a technological object (Latour 1996). As the Web makes it increasingly possible for those mundane objects which populate our surroundings to themselves leave traces, the concept of traceability is going from a research method to a business model for new startups in the Internet of Things. Yet, while these traceability interfaces offer to empower consumer activism, the reality is inevitably less dramatic than marketers claim. We are still left with a question of what to do with the knowledge that your local cheese is in fact a hybrid of Latvian cows, a German transportation company, and a Dutch producer. As depicted in the aforementioned television program, the locavore ideal does not necessarily create more choice, rather, it can paradoxically work to limit the choice to one. The core problems are what we talk about when we talk about context and who or what is this mythical end user in the life cycle of a sustainable product?

Problematizing Context Awareness

Over the course of the past decade media art projects such as MILK have explored how context-aware technologies might deepen engagement with place, for which the notion of locative media was developed (Tuters and Varnelis 2006, 357–363). While the field has become too diverse to summarize here, a unifying concept across the theory and practice of locative media is the notion of "context," which has relevance in discussing design approaches to food interaction. The convergence of mobile interfaces with location-aware technology constituted a new artistic media. By connecting decontextualized data to analog place, locative media seemed, rhetorically at least, to promise a recuperation of some of the lost aura associated with mechanical reproduction. Locative projects thus initially sought to explore the new possibilities opened up by this convergence in terms of connecting "real" and "virtual" space, a genre which could be considered as having culminated with the award of a Golden Nica for Interactive Art in 2003 to *Can You See Me Now*,[3] a media arts project which used GPS-enabled mobile technology to locate players within a video game that took place in urban space.

In the past half decade the concept of locative media has moved from the media arts fringe into the mainstream. Whereas locative artists were previously required to cobble their platforms together, by 2008 location awareness had become native on the iPhone making previously novel locative interfaces a relatively banal and commonplace feature. In recent years an entire ecology of location-aware apps has emerged for wayfinding, spatial annotation, and gamification as locative media has effectively become black-boxed into the new generation of mobile devices. This process can be seen to embody ANT's concept of the black-box, in reference to a stabilized actor network. The era of the invisible interface as anticipated by the future-oriented discourse of HCI research in the 1990s (Dourish and Bell 2011) had arrived, but rather than a future of computers embedding into the environment, it looked more like an "iPhone city" (Bratton 2009). Since the emergence of this media arts genre in the early 2000s, the technology that made it possible has become increasingly decentralizing and relational, moving from global positioning to what we might call object-oriented positioning as exemplified by the increasingly ubiquitous QR Code (which use Web-enabled devices' cameras in order to launch a URL). Applications designed to exploit this more relational or situated notion of location awareness can be considered as a first draft of the so-called Internet of Things (van Kranenburg 2008).

As a means of augmenting physical space and analog objects with data, locative media is usually conceptualized from the perspective of humans. This approach is, however, inconsistent with ANT's concept of generalized symmetry, which argues that all the elements in a network should be described in the same terms, ignoring any and all preexisting categories (object/subject, nature/culture, nonhuman/human) in favor of tracing relations within what they refer to as a network (Latour 1993, 103). While an entire discipline, namely human geography, is premised on the distinction between abstract space and embodied place, from the perspective of code, location is just another arbitrary value. While context in locative media is typically defined in rather absolute terms as geographic location, new standardized communications protocols make it possible for every single object on the planet to be part of the Internet of Things, thereby calling for a reassessment of the concept of context. If the goal of locative media is to bring context to information, there is no reason why it should remain wedded exclusively to location. From this perspective it is clearly the case that objects also communicate with us as well as with each other.

Context is not a location but a condition of possibility for a new network to appear, or, as Bruno Latour (2005, 186) reminds us, "context is what actors constantly do." What remains at issue however, is the position from which to measure. As location awareness continues to develop and become standardized into what we might call a locative apparatus, what emerges is a much more relational notion of proximity to objects. In addition to physically locating us in relation to them, objects become positioned in relation to one another, and crucially, through information visualization they become represented as gatherings of issues, in relation to which we can position ourselves. Through the careful work of representation every object could carry with it its own unique chains of reference, thereby revealing the substance of "the local" to be composed of an endless variation of scales.

Food, Technology, and Politics

Since 2007, a series of discussions around locative media took place in the Situated Technologies pamphlet series, which recently concluded with an exhibition and publication called *Sentient City* (Shepard 2011), the most discussed project of which was Trash Track. Like the MILK Project, Trash Track[4] traces and visualizes the path of consumer food objects. However, as opposed to the former (which told an origins story), Trash Track depicted the path of a project after it had been taken to the dump. Along similar lines consider the short polemical animated film titled *Story of Stuff*[5] about the lifecycle of consumer goods typically found in a department store. Originally produced in 2007, the film has been translated into fifteen languages and viewed by over twelve million people worldwide. Together these traceability projects can be understood within the broader context of a consumer movement toward increased transparency (Goleman 2009). Where early locative media art envisioned the emergence of alternative geographies (Tuters 2001), the current post-locative human-computer design applications have the potential to bring the controversial objects studied by actor network into the supermarket. As represented by this volume, a new field of HCI research is exploring the materiality of food as the bases of new "invisible" interfaces. Within this space our interest is to explore food interaction and discourse; specifically, what we have referred to above as metabolic interaction might be designed to spurn debate as to what gains entry into the individual and the collective body.

One of the few modern political philosophers for whom the stomach played a central role is John Dewey. In Dewey's conception the ideals of a liberal civic society "laid hold on the root of the character" through their repeated performance in terms of debate, discussion, and persuasion (Flammang 2009). His philosophy of education argued that urban civilization had disconnected children from the process of food production; thus, in his schools children were given the opportunity to make food together, a tradition which has been re-awoken a century later in the school farms movement spearheaded today by Alice Waters (2008). Dewey's concept of philosophical pragmatism, which has informed much of Latour's recent philosophical project, thus expanded out of a conception of a kind of civil engagement that emphasized learning by "doing," and indeed by cooking and eating.

In writing on food we see this kind of active engagement with civic politics everywhere, even among the seemingly opposed discourses of slow and local food versus functional food and biotechnologies. Indeed, the entire food movement in this sense reproduces the extreme positions in the environmental debates that envision salvation either through romantic return (deep ecology) or hypermodern control (eco-pragmatism). While the romantic Slow Food movement is tracing the origin of every food item in relation to tradition, geography, and political and economic structures, a functional "nutrigenomics" movement is tracing how ingredients interact with molecules inside our bodies. Seemingly philosophically opposed, both approaches nevertheless are faced with the design challenge of connecting data (provenance, molecules) with sensation (taste) and collectives built around various metabolic processes. The romantic advocates of small-scale local solutions like Vendana Shiva (2007) propose traditional farming as the solution to world hunger while eco-pragmatists like Stewart Brand (2009) advocate solutions involving genetically modified organisms (GMOs). While both consider themselves to be environmentalists concerned with global solutions, they approach the problem with nearly opposite solutions. The former emphasizes traditional tribal knowledge, the latter a technocratic fix. Both however are ultimately united by a common belief in things and more specifically food, as the proper site of deliberative politics. In the concluding section of this article on scenario design we will return to discuss how these various positions on food politics can be seen to relate to one another.

Product Ontology

In the global food industry, issues concerning food safety have brought about regulations which have in turn led to the widespread implementation of traceability in relation to supply-chain logistics. These regulations, combined with perceived consumer demand and pressure by consumers groups, have contributed to a trend toward corporate social responsibility that has encouraged manufacturers to analyze and redesign product life cycles (Goleman 2009). Until recently, these innovations had taken place on the supply side, signified to consumers via a variety of labeling schemes. A proliferation of marketing schemes and branding strategies oriented toward so-called lifestyles of health and sustainability (LOHAS), have combined to form a kind of discourse that Michael Pollan (2006, 134–140) refers to as "supermarket pastoral." Pollan has stated that after he wrote the best-seller called *The Omnivore's Dilemma: A Natural History of Four Meals* the question he was most frequently asked was, "What should I eat"? His replied by saying that it depended on one's concerns, whether they be animal welfare, the environment, health, labor, and so on (Pollan 2009).

Connecting products to the Internet of Things, via product barcodes and traceability applications, allows for direct connections between producers and consumers in order to trace and map an aspect of a product, such as its carbon footprint,[6] and to perform calculations on every single component of a given product by consulting innumerable sources of data categorized in terms of health, environment, and society. This concept has been referred to by one particular LOHAS marketer in terms of "product ontology" (Scelfo 2009). For example, the website GoodGuide is intended to enable consumers to retrieve ratings of products as well as make lists based on the filters—social, ecological, and/or health—that they prefer based on a database of over seventy thousand products with over eighty million bits of information.[7] Like many other similar start-ups, GoodGuide is looking to the next generation of the service that will rely more heavily on sensors (such as radio-frequency identification, or RFID) and automatically alert the shopper to the status of a product, for instance, or, in an alternative application type, analyze the shopper's credit-card purchases. Science journalist Daniel Goleman (2009, 174) captures the zeitgeist around the supposedly transformative powers of so-called ethical traceability, when he exuberantly claims: "If we get better, more complete information about the true effects of an item at the moment we are deciding whether to buy it, we could make wiser decisions. Such full disclosure can make each of us an agent for small, gradual changes that, when multiplied by millions, will ripple through the industrial enterprise, from manufacturing and design, through supply chains and transport, to the distant ends of consumption." While this claim might be criticized as technologically deterministic, and even naive in its faith in political change through consumer activism, it is worth looking more closely at the fundamental changes in the media ecosystem that motivate such claims.

In 1958 American free-market economist Leonard Read wrote an essay titled "I, Pencil" (1958/2008) in which he argued that since it was impossible to name all the antecedents that go into making a pencil and no single person therefore knew how to make one, the invisible hand of the market should therefore be left alone to do its work (Ridley 2010, 28). Arguably, however, this kind of neoliberal economic theory grew out of an era with different systems of measurement from the world of today. As marketing gurus Rachel Botsman and Roo Rogers (2010) argue, the growth in peer-to-peer technologies, as well as Internet-facilitated models for redistribution, is fostering a shift from ownership to access in which unused objects are increasingly being reconsidered in terms of their "idling capacity" (exemplary is the Air B&B service through which people rent out empty rooms in their homes). While they suggest the ongoing global recession as one factor driving this so-called rise of collaborative consumption, another factor appears to be a cultural shirt toward heightened ecological awareness. In this context William McDonough and Michael Braungart (2002) have developed the notion of a 360-degree loop in which the analysis of a product's life cycle is incorporated into its design such that manufacturers consider their waste products as resources to be fed back into the system. What they call cradle-to-cradle design can be understood as both a critique of industrial design practice as well as a proscription for sustainable design in a closed ecosystem. Rejecting the 3 Rs of environmentalism (reduce, reuse, and recycle) as insufficiently revolutionary, their approach effectively transforms planned obsolescence from a sin into a strategy. McDonough and Braungart's notion of "cradle-to-cradle" accounts for both the provenance and the consequence of their designs. Cradle-to-cradle objects are thus conceived as merely temporary material instances of a larger metabolic process. From the gene-disrupting capacity of certain plastics to the global trade in e-waste, the cradle-to-cradle designer is concerned not only with the objects, but also with all its various externalities. As opposed to Read's neoliberal economic theory, which decried any form of planning as impossible due to the complexities of these systems, the notion here is that as the tools of measurement change so too should the economic theories. To that end, design theorist Bruce Sterling (2005) has claimed that traceability is in fact making the invisible hand visible by connecting commodities with metrics that allow for objects to represent their conditions of production.

In his 2005 design manifesto, Sterling envisioned a "wand" that would make it possible to connect objects with their dataclouds through a number of technologies including RFID and what he calls "geo-locative media," in order to awaken consumers to the myriad negative externalities produced by current unsustainable production techniques. Similarly, Latour (2004a) has considered the role of advanced technological apparatus at the level of the senses in a discussion of the *mallette à odeurs* used by professional perfume smellers to sensitize their noses. He offers this device as a metaphor for a material, and highly artificial, practice through which the human body learns to register and become sensitive to what the world is made of. This capacity of cultural objects and their communities of practices to cybernetically steer politics is likewise recognized by the Slow Food movement, which seeks to promote cultural preservation through taste—although, in distinction to the discussion thus far, this movement embraces traditional food practices, turning its back on science technology innovation. Participants in Slow Food "taste workshops" are, for instance, exposed to heritage foods in order to awaken their palates, which Slow Food claims, in Marxist terms, have become de-skilled through constant exposure to flavorless industrial food commodities. Intriguingly, Slow Food thus seeks a political awakening not through communicative reason but by appealing to taste, in terms of olfaction (or, as their critics would claim, in terms of distinction).

HCI applied to food may fundamentally transform the experience of eating and shopping. For example, HCI can provide us tools through which to become sensitized to the ecological impacts of eating choices. It seems entirely likely that traceability technology will soon be used to augment package labeling, for instance, to bolster the claims of LOHAS certification schemas like Fairtrade. It is not so hard to potential value-add for the ethical consumer interested in the conditions of production of her coffee. Given how easily manipulated people's olfactory taste has been shown to be (Trillin 2002), we can conceive how such apparatus could literally enhance a food's very flavor. Such a system is likely to rely on a user's Internet-enabled mobile device; it remains to be seen, however, how well mobile technology and food fit together in practice, and whether in the end such a technology would lessen or increase the cognitive load. Whatever the case, while we may envision a more fair and sustainable world facilitated by metabolic interaction, it would behoove us to bear in mind Paul Dourish's critique of HCI as an anti-politics machine (2010), in reference to its commitments to conceptions of technological progress and determinism as well as its neoliberal conception of the individual "user" as the ultimate arbiter of social change. His proposed solution is to design a social network that would allow the green consumer to see an object in terms of matters-of-concern, manifesting communal forms of agency, with the ultimate goal of effecting change at the level of state regulation: "the action you are about to take aligns you with X but against Y" (Dourish 2010), in other words through agonistic politics (Schmitt 2007). Although Coca-Cola now offers an online service through which you can trace the origins of your Coke,[8] we can be reasonably certain that a corporation will not be responsible for developing the disruptive technology called for by Dourish.

Cosmopolitics of the Table

How might metabolic interaction design facilitate this agonistic food politics while avoiding the naive and ridiculous situation we encountered by the two diners at the outset of this essay? Perhaps a good place to start is with Jean Anthelme Brillat-Savarin, who "founded the genre of the gastronomic essay" (Mennell 1996, 267). As we will try to show, he is also the first metabolic interaction designer. In his famed "The Physiology of Taste," Brillat-Savarin's second Aphorism states that "beasts feed: man eats: the man of intellect alone knows how to eat" (2004, 3). This insight is supported by the famous historian of science, Sir G.E.R. Lloyd, who argues that since antiquity eating has been the basis upon which we define ontological distinctions between beasts, humans, and gods.[9] Brillat-Savarin's original insight and its older Greek resonance still hold true for today's extreme food practices and the view of humanity they present.

Eating is as much a social, cultural, and political act as it is a natural act, happening on a table as a place where discourse, nature, and culture meet to connect our animality with what defines our civilization. Brillat-Savarin (2004, 37–38) considered eating together as the original and most authentic form of public participation in political and social structures: "no great event, and we make no exception even of conspiracies, has ever come to pass, which was not first conceived, worked out, and set in train over the festive board." To know how to eat, for Brillat-Savarin, involves not only techniques of cooking but more importantly "pleasures of the table" and gourmandism that position gastronomy in the context of sciences, philosophy, and even politics and business. Savages would decide questions of peace and war in the midst of their feasts, and villagers conduct their business at the inn, but for Brillat-Savarin, the modern "political gastronome" organizes the whole globalized world like a "well-spread table." "Gastronomy takes note of men and things, to the end that essential knowledge, may pass from one country to another; and thus it comes about that a well-spread table seems an epitome of the world, every part of which is duly represented" (Brillat-Savarin

2004, 36). His "well-spread table" thus represents the hybrid world of men and things in which individual actors meet to define their alliances and networks based on metabolism which connects and transforms elements into new units. Even when considered from Brillat-Savarin's nineteenth-century perspective, it is odd that HCI has been so slow to consider food as its object of research when it seems to provide such an ideal model for assembling humans and nonhumans, society and nature. Following Brillat-Savarin, the goal of our design scenario, as discussed below, is therefore to bring together the rational and the instinctual in new equilibria, both within us individually and in various forms of collectives.

Brillat-Savarin's gastronomical meditations share similar goals with the various performative methodologies in art and critical design discussed above and bear a striking conceptual relation to concepts of actor-network theory. Brillat-Savarin's descriptions of "banquets" embodies what Latour (2004b) refers to as "cosmopolitics," in which we see ourselves as enmeshed in heterogenous networks of human and nonhumans. Every banquet or dinner is not only an opportunity to feed ourselves but to interact with various people and their interests. The purpose of gourmandism for Brillat-Savarin (2004, 111) is that it "draws out that spirit of fellowship which daily brings all sorts together, moulds them into a single whole, sets them talking, and rounds off the angles of conventional inequality." What drives his research is not a pretense of authority but a "praiseworthy spirit of curiosity," an interest in trends or "fear of being behind times" (2004, 9), and a pragmatists concern with things and practices: "Considered in its relation to political economy, gourmandism is the common bond which unites the nations of the world, through the reciprocal exchange of objects serving for daily consumption . . . [and] causes wines, spirits, sugar, spices . . . to cross the earth from pole to pole" (107). What fascinates Brillat-Savarin is what fascinates us, the capacity for food to operate at various scales and levels: from body to society, from history to everyday life, from past to present.

To return then, to the opening anecdote regarding the locavore nostalgic for a lost sense of "the local," our approach thus far has been to deconstruct notions of the user and her context by tracing the metabolic pathways of global food logistics from farm to fork. Food is an object and a commodity that connects us on a social and political level with the food industry and various policy regulators. At the same time, food is also an energy source and a building block for our cells that connects us to the planet on a molecular level. Crucially, food interaction must be appreciated as a set of activities both pleasurable and guilty that connect us with other living beings in intensive emotional and embodied exchanges. Having now discussed the well-spread table as a kind of political assembly, in this second section we go on to look at some design experiments based on the metaphor of tracing metabolic interactions, the objective of which have been to design scalable interfaces that connect issues relating to the planet with issues relating to markets with issues relating to our bodies.

Design Scenario: Diet-Tribe Diatribes

We demonstrate to anyone whose soul has fallen to pieces that he can rearrange these pieces of a previous self in what order he pleases, and so attain to an endless multiplicity of moves in the game of life. As the playwright shapes a drama from a handful of characters, so do we from the pieces of the disintegrated self build up ever new groups, with ever new interplay and suspense, and new situations that are eternally inexhaustible.

Herman Hesse ([1927] 2004, 174)

In our design research we take seriously the ontological implications of new technologies and scientific "discoveries" in terms of how they might impact conceptions of human subjectivity and community. To this end our creative response builds on the imagery of the post-human collective from philosophy and from literature. When we trace and transform food flows we work toward the convergence of the human and what Gilles Deleuze and Felix Guattarri (1987) refer to as the machinic phyla in which every new actor that is metabolized by the body, the social system, or some other aggregate opens the opportunity for radically new forms of community and materialist political action. As Deleuze (in Marks 1998, 27) states, "these days it's often said that man is confronting new forces: silicon and no longer carbon, the cosmos rather than the world." Although most pervasive networked technologies deal with nonintrusive objects outside of our bodies, we are interested in how scientific insights and silicone innovations enter open our viscera onto a new cosmos.

In a series of projects that we developed in 2010 and 2011 (Kera and Tuters 2011), we sought to explore this post-human hypothesis through HCI designs which themselves assumed a generalized symmetry in regard to agency across all elements in the network (in this case between humans, leftovers, refrigerators, and DNA molecules). We created prototypes ("FridgeMatch"[10]), performed scenarios ("Eat What You Are: Dining on Biopolitics"[11]) and explored metaphysical problems relating to concepts of post-human agency inspired by philosophy and speculative fiction. We consider the resulting assemblages as both instances of public deliberation on emergent scientific fields or issues (sensor networks, the Internet of Things and leftovers, nutrigenomics), and as philosophical probes into thinking about future communities, forms of co-existence, and indeed co-evolution, beyond narrow and, ultimately, unsustainable conceptions of merely a human user. Neither interested in user studies nor in facilitating scientific policy discussions, our approach to scenario design might perhaps best be considered as experimental, performative, and ultimately philosophical.

We were initially inspired by Margaret Atwood's *The Year of the Flood* (2009) to speculate on technosocieties in which identities are formed around extreme food practices. We began by creating a speculative narrative reflecting on contemporary food identity politics and technology in which four food *diet-tribes* were defined by

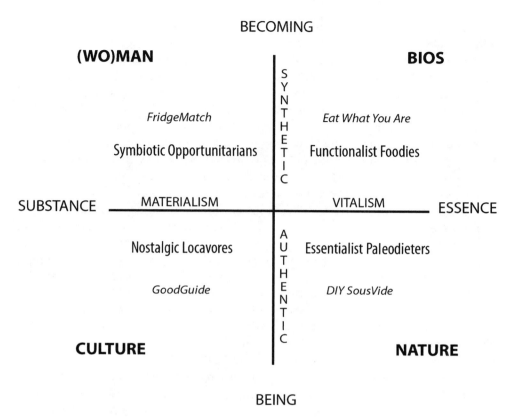

Figure 14.1
Diagram for "Diet-Tribe Diatribes" design scenario

opposing views, or *diatribes*. We in turn extrapolated a critique from these sources in order to design evocative prototypes which, following the approach of critical design, sought "to challenge narrow assumptions, preconceptions and givens about the role products play in everyday life" (Dunne and Raby, n.d.). Each diet-tribe corresponds with a position in the diagram for the Diet-Tribes Diatribes design scenario (figure 14.1). The y-axis is concerned with ontology—with the positions on the bottom of the graph oriented toward notion of authentic being—in contrast to the positions at the top that are concerned with a synthetic concept of becoming. The x-axis is concerned with metaphysics, with positions on the left corresponding to materialist substance and those on the right to vitalist essence. In addition, the graph sets "Nature" and "Culture" in binary opposition to "(Wo)man" and "Bios." The objective of this exercise was to locate design projects—what we have thus far called metabolic interaction—in each of the four quadrants.

Based on the object-oriented approach to politics discussed in relation to Latour and Brillat-Savarin, we were interested in how new forms of community could be observed as forming around new practices. So far this article has explored how techniques for locative traceability might be used to address questions of food provenance and product ontology. Our scenario identified these concerns with a group we called the Nostalgic Locavores, associated in the diagram with the "Culture" quadrant. Obsessed by a notion of authenticity, this diet-tribe was connected to the historical preservationist diatribes of the Slow Food movement. By contrast, the group that we referred to as the Essentialist Paleodieters were characterized by their faith in natural science theories concerning the authentic origins of man as a predatory animal. The Paleodieters thus reject the Locavores' romantic ideals rooted, as they are, in the community of man, in favor of diatribes from evolutionary biology that advocate the health benefits of the meat-based diets of our violent ancestors. Grounded in this faith, members of this community are preoccupied with preserving the "essential nutrients" of food for which they have developed high-tech cooking techniques.[12] Accordingly, one of our first projects involved engaging in participatory design charrettes with members of this diet-tribe, by building, among other things, a DIY SousVide cooking appliance that used a vacuum to cook at a very low temperature (Kera and Tuters 2011).

Critical design has developed the notion of the design prop as an object overflowing with reference to a diegetic world beyond itself, through which "the viewer can be drawn into the conceptual space of the object in use rather than an appreciation of the thing in itself" (Dunne 1999). The goal of such design props is not necessarily to envision a future, open discussion but also question our philosophical assumptions. Through creating these design props we reflect upon the future of metabolic interactions across scales in order to ask philosophical and ethical questions about nonhuman and post-human futures. While we have considered how the diatribes of Slow Food are based on notions of authentic historical communities, our design scenario suggested alternative "synthetic" forms of community engagement based on modern life styles, from dating to self-surveillance, rather than on the enshrinement of traditions. As opposed to the Nostalgic Locavores, concerned as they were with grandiose notions of cultural patrimony, the diet tribe that our scenario referred to as the Symbiotic Opportunitarians were future oriented yet tactical by nature, less concerned with either the authentic qualities of food or their essential nutrients than with food's role as a catalyst for emergent forms of community in which being is conceived rather as a process of becoming (what Deleuze and Guattari [1987, 232–309] refer to as "becoming minor," "becoming woman," or "becoming animal").

As the design prop that we created for the Symbiotic Opportunitarian, "FoodMatch" allowed strangers to share the leftovers in their fridges via Facebook. In practice "Food-Match" worked like a dating website in which your leftover expressed your personality. "FoodMatch" was thus designed to accommodate a lifestyle based on eating only leftover food, inspired by the parasitic practices of dumpster-diving "freegans" whom animal-rights activist and utilitarian philosopher Peter Singer (Singer and Mason 2006) refers to as the "ultimate" in ethical eaters. Like the freegans who build their tactical food politics in the interstices of supermarket back alleys, we sought to exploit the affordance of Facebook's identity supermarket. Participants in FoodMatch uploaded their lists of leftovers to the site: the leftovers, these sad reminders of lonely atomized lives, thus became a means of social interaction—when possible, "FoodMatch" consulted matched users by location inviting them to cook together in person. As such, "FoodMatch" used the very substance of food as a catalyzing agent around which new types of communities and perhaps even new sexual identities might emerge.

Having discussed the metabolic interface initially at the global/ecological and subsequently the social/intersubjective scales, we want to conclude here with a project that takes the cult of traceability inside the individual human body. In May of 2011, we staged "Eat What You Are: Dining on Biopolitics," a performance at the brmlab.cz hackerspace in Prague in which meals were specifically tailored to phenotypes as captured in the diner's 23andMe genetic test results. (The blog site 23andMe offers gene-related information regarding ancestry, health, and scientific research.) Building on the "functional food" hypothesis, meals were designed to appeal to the appetites of the diner's genes rather than her own preferences. Dishes were thus accompanied by extremely specific descriptions. For example, one such description explained that 70 percent of the food on a diner's plate (stuffed mushrooms) referred to her dominant Ashkenazim ancestry, while 6 percent of the food (hummus) referred to some unclear Middle Eastern genetic heritage, and that the 6 chromosome—which was the most clearly Middle Eastern part of this diner's phenotype—played an important role in both immune response and sexual attraction as the base for genes closely linked to olfactory receptors. This project was thus intended to problematize the concept of the user/eater by introducing the "selfish gene" (Dawkins 1976) as a kind of Cartesian evil demon.

"Eat What You Are: Dining on Biopolitics" sought to trace emergent forms of metabolic exchange at the phenotypic scale, as made possible by personalized DNA profiles. Our objective with this performance was to reimagine the experience of eating for the coming age of corporate-enabled personal genomics, a period anticipated by Michel Foucault (2010) in terms of the management of life in all its forms. The project thus corresponded with the "Bios" quadrant in our diagram, which refers to the Greek word for "life" as well as to the Basic Input/Output System in computers. The diet tribe we associated with this quadrant was the Functionalist Foodies, whom we modeled on the movement to measure and quantify all input and outputs from the body as championed among others by the Silicon Valley productivity guru Tim Ferriss (2010). Similar to the Essentialist Paleodieters with their faith in nutrients, the Functionalist Foodies thus related to food in terms of invisible vital force as revealed by natural science.

Ultimately all four of our diet tribes approached food in terms of political assemblies. In each case then the technologies and design props discussed above were intended to demonstrate how the metaphor of metabolism can be used to problematize certain assumed notions in both food politics and HCI, but also to new aggregates which we have framed in political philosophical terms as *cosmopolitcs* and *biopolitics*. As opposed to the language of militancy (within which the concept of biopolitics is so often deployed), our objective with the projects generated by the Diet-Tribe Diatribes design scenario can be understood as staging an exchange between a friendly gathering of strangers (humans, food, molecules, ideologies) around Brillat-Savarin's "well-spread table." Rather than evoking the notion of exodus from the biopolitical to some holy site (Virno and Hardt 1996), as discussed in the opening anecdote, this concept of metabolic interactions seeks to encouraged eaters to deliberate on what gained access to the body politic, where the concept of the body is in a process of becoming.

Summary

We have developed the metaphor of metabolism for interfaces that connect and aggregate collectives across a variety of scales. We have framed this work in terms of ontological and political questions around the concept of the user and her context in both food politics and HCI. We have suggested that data can enhance appetite and turn the experience of eating into political and social mobilization around the table. We have framed new technological innovations and scientific insights discussed in terms of post-human theories of agency and ontology, developing the Diet-Tribe Diatribes design scenario as implementations of these insights, out of which we generated a number of prototypes; we include the recipe from one of these at the end of this chapter. We have drawn upon these analyses to consider food interaction by tracing metabolic exchanges across multiple scales in order both to critique food politics notion of "the local" and to contemplate new forms of collectivities. In so doing we have also suggested a critique of HCI concepts of the user and of context. Our design research has sought to create new systems and interconnections between units (genes, ingredients) and wholes (bodies, society) through the concept of metabolic interaction by which design thinking might engage in the creation of new biopolitical collectives. Rather than existing as a retrenchment in the safety of the outdated ontological categories of user and context, our work has sought to multiply the potential positions for action from farm to fork to genetic phenotype in which metabolic interaction serves as a metaphor for the political dimensions of pleasure and taste. Although governmentality theorists argue that empire exercises control through the management of life itself (Hardt and Negri 2001, 22–41), our gambit has been that critical designers can be imagining new collectivities as tactical biopolitics.

ADRA2A & MTHFR LAMB WITH SPINACH & POTATOES

This recipe pairing was developed by the authors in collaboration with Alexandra Spencer and Meng Weng Wong for the "Eat What You Are: Dining on Biopolitics" project on May 2011 at the brmlab.cz hackerspace in Prague. The method of cooking, using a sous vide (French, "under vacuum"), a technique in which the food cooks in a vacuum-sealed pouch, submerged in a water bath, at a precise temperature to preserve flavor and nutritional value.

Actors and Ingredients

Homozygous and heterozygous individuals with rs1800544 (C;G) (C;C) for the human adrenergic Alpha-2A receptor (ADRA2A) gene.

Homozygous individual with rs1801133(T;T) genotype of the methylenetetrahydrofolate reductase gene (MTHFR).

Milk-fed lamb, Yield Grade 1: based on USDA lamb quality and yield-grade standards determined on the basis of the adjusted fat thickness over the rib eye muscle between the 12th and 13th ribs (Yield Grade 1; 0.00 to 0.15 inch).

New Zealand organic spinach, ranked 10 in the "Health Impacts" and "Product Management" categories by GoodGuides.com with certification by the USDA.

Russet organic potatoes, ranked 10 in the "Health Impacts" and "Product Management" categories by GoodGuides.com with certification by the USDA.

Preparation

1. Vacuum-seal 1 lamb shoulder seasoned with rosemary, cinnamon, mint, paprika, thyme, parsley, black pepper, crushed garlic, and sea salt. Place in sous-vide appliance for 24 hours at 135°F (58°C).
2. At the 23-hour mark, preheat oven to 400°F (200°C). On bottom rack of oven, bake the potatoes, rubbed with olive oil and sea salt, for 1 hour.
3. Carefully remove lamb from sous-vide bag and place in oven on top rack until fragrant and fat cap is browned (about 10 to 20 minutes).
4. In large skillet or wok, heat olive oil on medium-high. Add one clove of thinly sliced garlic to season oil, stir and remove from oil. Increase heat and add rinsed spinach, tossing leaves until all are wilted but still bright green. Place cooked spinach in sieve to drain excess water, and season with lemon juice, sea salt, and olive oil.

Menu Description for Homozygous Individual with rs1801133(T;T) Genotype

Lamb for the main course is a healthy source of vitamin B12 and B3 and is lower in saturated fat, supporting the balanced function of both ADRA2A and MTHFR genes. Since these individuals have good sugar metabolism and low risk of diabetes, we include potatoes as a side dish, but meals will be served in portions of different sizes depending on the individual sugar intake efficiency status (ADRA2A gene) with the right proportion of green veggies like spinach, which balances the individual's need for folates (MTHFR gene). This person with the MYHFR gene will get more spinach as a side dish because of the risky TT variant on the rs1801133 SNP related to the MTHFR gene regulation of folates. This genotype (10 percent of the population) has only 30 percent of the expected MTHFR enzyme activity related to folates when compared with the most common genotype, (C;C). Folates play an important role in the complete development of red blood cells, which carry oxygen around the body and help maintain healthy circulation of the blood throughout the body by preventing build-up of homocysteine. Folate status is also critical for normalizing antibody production and detoxifying the histamine the body produces as part of allergic reactions. It regulates all the detox enzymes that determine how well a person clears toxins like heavy metals. Hence folates play a crucial role in the meal.

Notes

1. http://www.ifc.com/videos/portlandia-is-it-local.php.

2. http://milkproject.net.

3. http://www.blasttheory.co.uk/bt/work_cysmn.html.

4. http://senseable.mit.edu/trashtrack.

5. http://www.storyofstuff.com.

6. Sourcemap—Open supply chains & carbon footprint, http://www.sourcemap.com.

7. GoodGuide—Methodology overview, http://www.goodguide.com/about/methodologies.

8. See Trace Your Coke online at http://www.cokecorporateresponsibility.co.uk/journey-of-a-coke/trace-your-coke.aspx.

9. http://backdoorbroadcasting.net/2010/09/sir-geoffrey-lloyd-humanity-between-gods-and-beasts-ontologies-in-question.

10. http://www.fridgematch.org.

11. http://brmlab.cz/event/denisa_kera.

12. See, for example, SousVide mailing list at http://www.modernpaleo.com/sousvide.html.

References

Antonelli, Paola. 2011. *Talk to me: Design and the communication between people and objects*. New York: The Museum of Modern Art, New York.

Atwood, Margaret. 2009. *The year of the flood: A novel*. New York: Nan A. Talese/Doublebay.

Botsman, Rachel, and Roo Rogers. 2010. *What's mine is yours: The rise of collaborative consumption*. New York: HarperBusiness.

Brillat-Savarin, Jean Anthelme. 2004. *The physiology of taste*. Whitefish, MT: Kessinger Publishing.

Brand, Stewart. 2009. *Whole earth discipline: An ecopragmatist manifesto*. New York: Viking Adult.

Bratton, Benjamin H. 2009. iPhone city in digital cities. *Digital Cities AD: Architectural Design*, ed. Neil Leach. Malden, MA: John Wiley & Sons Ltd.

Dawkins, Richard. 1976. *The selfish gene*. New York: Oxford University Press.

Deleuze, Gilles, and Felix Guattari. 1987. *A thousand plateaus: Capitalism and schizophrenia*. Minneapolis: University of Minnesota Press.

Dourish, Paul. 2010. HCI and environmental sustainability: The politics of design and the design of politics. In *DIS 2010: Proceedings of the 8th ACM Conference on Designing Interactive Systems*, 1–10. New York: ACM.

Dourish, Paul, and Genevieve Bell. 2011. *Divining a digital future: Mess and mythology in ubiquitous computing*. Cambridge, MA: MIT Press.

Dunne, Anthony. 1999. Hertzian tales: Electronic products, aesthetic experience and critical design. London: Royal College of Art Computer Related Design Research Studio.

Dunne, Anthony, and Fiona Raby. n.d. Critical design FAQ. *Dunne and Raby*. http://www.dunneandraby.co.uk/content/bydandr/13/0.

Ferriss, Timothy. 2010. *The 4-hour body: An uncommon guide to rapid fat-loss, incredible sex, and becoming superhuman*. New York: Crown Archetype.

Flammang, Janet A. 2009. *The taste for civilization: Food, politics, and civil society*. Chicago: University of Illinois Press.

Foucault, Michel. 2010. *Birth of biopolitics: Lectures at the College de France*. New York: Palgrave.

Goleman, Daniel. 2009. *Ecological intelligence: How knowing the hidden impacts of what we buy can change everything*. New York: Broadway Books.

Gordon, Eric, and Adriana de Souza e Silva. 2011. *Net locality: Why location matters in a networked world*. Malden, MA: John Wiley & Sons Ltd.

Hardt, Michael, and Antonio Negri. 2001. *Empire*. Cambridge, MA: Harvard University Press.

Harris, Edmund. 2009. Neoliberal subjectivities or a politics of the possible? Reading for difference in alternative food networks. *Area* 41 (1):55.

Hesse, Herman. 1990. *Steppenwolf: A novel*. New York: Macmillan Holt Paperbacks.

Kera, Denisa, and Marc Tuters. 2011. *Social stomach: Performative food prototypes*. Paper presented at Creative Science 2011, July 25–26, Nottingham, England.

Latour, Bruno, and Peter Weibel, eds. 2005. *Making things public: Atmospheres of democracy*. Cambridge: MIT Press.

Latour, Bruno. 1993. *We have never been modern*. Cambridge: Harvard University Press.

Latour, Bruno. 2004a. How to talk about the body? The normative dimension of science studies. *Body & Society* 10 (2–3):205–229.

Latour, Bruno. 2004b. *Politics of nature: How to bring the sciences into democracy*. Cambridge: Harvard University Press.

Latour, Bruno. 2005. *Reassembling the social: An introduction to actor network theory*. Oxford: Oxford University Press.

Latour, Bruno. n.d. *EXPOSITIONS | EXHIBITIONS: Dingpolitik-Making things public*. http://www.bruno-latour.fr/node/333.

Latour, Bruno. 1996. *Aramis, or, The love of technology*. Cambridge: Harvard University Press.

Read, Leonard E. 1958/2008. *I, pencil: My family tree as told to Leonard E. Read*. New York: Foundations for Economic Education.

Marks, John. 1998. *Gilles Deleuze: Vitalism and multiplicity*. London: Pluto Press.

McDonough, William, and Michael Braungart. 2002. *Cradle to cradle: Remaking the way we make things*. New York: North Point Press.

Mennell, Stephen. 1996. *All manners of food: eating and taste in England and France from the Middle Ages to the present*. 2nd ed. Champaign: University of Illinois Press.

Pollan, Michael. 2006. *The omnivore's dilemma: A natural history of four meals*. New York: Penguin.

Pollan, Michael. 2009. Deep agriculture. *Seminars about long-term thinking*, Tuesday May 5. http://longnow.org/seminars/02009/may/05/deep-agriculture.

Ridley, Matt. 2010. *The rational optimist: How prosperity evolves*. New York: Harper.

Scelfo, Julie. 2009. Apply these product ratings generously. *New York Times,* May 20. http://www.nytimes.com/2009/05/21/garden/21qa.html.

Schmitt, Carl. 2007. *The concept of the political*. Chicago: University Of Chicago Press.

Shepard, Mark. 2011. *Sentient city: Ubiquitous computing, architecture, and the future of urban space*. Cambridge: The Architectural League/MIT Press.

Shiva, Vandana, ed. 2007. *Manifestos on the future of food and seed*. London: South End Press.

Singer, Peter, and Jim Mason. 2006. *The way we eat: Why our food choices matter*. New York: Rodale Books.

Sterling, Bruce. 2005. *Shaping things*. Cambridge: MIT Press.

Trillin, Calvin. 2002. The red and the white: Is it possible that wine connoisseurs can't tell them apart? *New Yorker*, August 19. http://www.newyorker.com/archive/2002/08/19/020819fa_fact.

Tuters, Marc. 2001. Variation on a videogame or spatial graffiti: The socio-spatial and futurological implications of augmented reality and location awareness. In *Proceedings of the Seventh International Conference on Virtual Systems and Multimedia*, 517–526. Washington, DC: IEEE.

Tuters, Marc, and Kazys Varnelis. 2006. Beyond locative media: Giving shape to the Internet of things. *Leonardo* 39 (4):357–363.

van Kranenburg, Rob. 2008. *The Internet of things: A critique of ambient technology and the all-seeing network of RFID*. Amsterdam: Institute of Network Cultures.

Virno, Paolo, and Michael Hardt, eds. 1996. *Radical thought in Italy: A potential politics*. Minneapolis: University of Minnesota Press.

Waters, Alice. 2008. *Edible schoolyard: A universal idea*. San Francisco: Chronicle Books.

15 Food Futures: Three Provocations to Challenge HCI Interventions

Greg Hearn and David Lindsay Wright

What images of the future might we invoke in examining the futures of food? What do we mean by "an image of the future," and how has food figured in imagining the future? A comprehensive discussion of food futures should focus on many different aspects of the greater food industry: geopolitics, agriculture, manufacturing, handling, transportation, retailing, and security. This palette of possibility is beyond our scope however. We take our specific point of departure from the intention of this book, namely, to consider the rise of Web 2.0 technologies and whether they enable an increased potential for a culture of participation and creativity in relation to food futures. Drawing on the idea of communicative ecologies (Foth and Hearn 2007) as a way to explain the interaction of social, discursive, and technological factors in human action, we introduce the idea of "food zones" as a tool for thinking about and understanding food in society. We suggest that the idea of food zones offers a useful language and point of intervention to positively transform food practices and cultures. We then develop a taxonomy we call the Food Futures Cube to depict some possible points of intervention in the overall food production and consumption systems into the future. A comprehensive consideration of the Food Futures Cube is beyond our scope here. Our more modest objective is to simply provoke novel ways of looking at the future possibilities for food that might be addressed by future human-computer interaction (HCI) interventions.

To do this, we construct flash fiction–type scenarios about food futures. The term "flash fiction" has numerous variations, but for our purposes we use it to denote a very compact and to-the-point "story" or vignette, in a science/speculative fiction style, about food futures. As for the subingredients that go into creating our flash fiction–type future provocation scenarios, we deliberately borrow from a wide variety of sources including a diverse film and video clip line-up, science fiction literature, and scholarly literature. With this eclectic "mixed pizza" approach to the theme of food futures, our ultimate goal is not to usher in a definitive voice on the topic but rather to offer a varied menu of "future provocations" about food which we hope will simultaneously signal alarm bells, offer unique insights, and give rise to novel combinations of possible, probable, and preferable futures.

The Image of the Future

To discuss the future of food necessitates we initially consider our idea of "the future" more broadly. In his seminal work, Fred Polak (1973) introduces the image of the future within society and through history.

The human capacity to create mental images of the "totally other"—that which has never been experienced or recorded—is the key dynamic of history. At every level of awareness, from the individual to the macro-societal, imagery is continuously generated about the "not-yet." Such imagery inspires our intentions, which then move us purposefully forward. Through their daily choices of action, individuals, families, enterprises, communities, and nations move toward what they imagine to be a desirable tomorrow. (Boulding 1988, 19)

Utopian thinking is one version of this futures image making.

Utopias elude definition. The genre merges, at its edges, into related forms: the imaginary voyage, the earthly or heavenly paradise, the political manifesto or constitution. But an average, middle-of-the-road utopia will include transit to some other place, remote in space or time or both, where the inhabitants are different from us, perhaps recognizably human, perhaps not, and where something can be learned about how life should be lived. (Carey 1999, 1)

Polak (1973) devoted over seventy pages to the roles and functioning of utopias in history and upon the future.

Many utopian themes, arising in fantasy, find their way in to reality. Scientific management, full employment, and social security were all once figments of a utopia-writer's imagination. So were parliamentary democracy, universal suffrage, planning, and the trade union movement. The tremendous concern for childrearing and universal education, for eugenics, and for garden cities all emanated from the utopia. (Polak 1973, 137)

For Polak, the utopia as "image of the future" constitutes one of the most powerful forces in shaping history, and, by implication, futures, through the confluence of three possible roles: as a buffer, as a driving force, and as a trigger of social progress.

Inspired by William Morris's great utopian novel, *News from Nowhere*, Stephen Coleman and Paddy O'Sullivan comment:

Let us imagine that life is not as it is, but as it might one day be. Let us inspect the unknown terrain of the future as if we were about to inhabit it. . . . The imagined future is a subversive force: the more who imagine a different kind of future, and imagine constructively, materially, and determinedly, the more dangerous utopian dreams become. They grow from dreams to aims. (1990, 10)

Two Images of the Future of Food

Food is a feature of many utopian visions such as the archetypal utopia, the *Land of Cockaigne*. As Cockaigne researcher Herman Pleij (2001) notes:

Everyone living at the end of the Middle Ages had heard of Cockaigne at one time or another. It was a country, tucked away in some remote corner of the globe, where ideal living conditions prevailed: ideal, that is, according to late-medieval notions, and perhaps not even those of everyone living at the time. Work was forbidden, for one thing, and food and drink appeared spontaneously in the form of grilled fish, roast geese, and rivers of wine. One only had to open one's mouth, and all that delicious food practically jumped inside. One could even reside in meat, fish, game, fowl, or pastry, for another feature of Cockaigne was its edible architecture. (3)

Charles Leadbeater (2002) suggests how Western life in the Information Age is in fact a contemporaneous living out, a realization, of medieval dreams as evidenced in prototype form in the Cockaigne stories; the modern consumer society in many ways is living out a medieval myth: fantastical stories about pleasures of life in the Land of Cockaigne set out in the *Sterfbeok* (*The Book of Death*) written in 1491. Cockaigne was replete with a never-ending supply of food, and freedom from disease, bad weather, and other sources of fear and uncertainty. It is from these prototypical utopian stories that much of our modern consumer culture unwittingly seeks to realize the medieval Cockaigne idyll as a contemporary "Technotopia." Leadbeater observes that "innovation-driven societies constantly supply greener grass for the other side of the valley" (2002, 168).

In contrast to these technotopian visions, Japanese scholar Shinichi Tsuji (2001) argues in his book *Slow Is Beautiful* that contemporary life, not just in Japan but all technologically developed cultures, is overly speed-dependent and continually accelerating to the detriment of human relationships with their respective food cultures. To counteract this, Tsuji established the Sloth Club and more recently has been involved with a number of environmental and sustainability friendly organizations such as the company Slow. Tsuji's groundbreaking work in slow living has inspired numerous spin-off ventures such as the Slow Life Café in Hakodate. *Slow is Beautiful* presents a powerfully subversive critique of Japanese conceptions of time, investigating how the politics of temporality impinge upon all areas of Japanese life, and by extension, contemporary global culture.

The costs and effects of the relentless pursuit of the accelerating society is concisely summarized by Japan commentator Patrick Smith (1998, 173):

Modern Japan has always been obsessed with speed. We can attribute to a fundamental sense of urgency many of the mistakes it has made over the past century, including its decision to erase local identity rather than incorporate it. The frenetic pace set by modern Japanese leaders, to make the matter clear, has never had anything to do with culture, tradition, or innate character traits. It began with the desire to *catch up*, which reflected anxiety, felt inferiority, and fear. (italics added)

Returning to Tsuji's *Slow Life*, one of his key aims is to demonstrate how a slower life is in the long term not only a more human way to live, but simultaneously advantageous to our natural global environment. But there is a core *myth* that Tsuji deconstructs: the myth of laziness which claims that time equals money, and the normalized assumption that time, power, and efficiency exercised via control and manipulation function as *the* key to the competitive advantage of prosperous nations such as Japan.

Tsuji counteracts this often taken-for-granted assumption by substituting his own myth, the myth of *natural* time: a return to simpler times when the notion of time itself was more simply construed. According to Tsuji, this will give twenty-first-century humans an augmented perception of self-control over their lives and bodies, whose outcomes would include giving time back to the natural world. The cynic might be inclined to argue that a slow future neglects the fact that the emerging social media generations have constructed their own conceptions for space and time, and that slow life is nothing more than a philosophy of nostalgia oblivious to the needs and realities of accelerating social mobility.

Despite this, in an age of increasing concern for environmental sustainability, including human sustainability, the pace of life and the pace of consumption are important sites of debate that cannot be ignored. In fact, we suggest that the future of food will be largely influenced by debates around the future of consumption, and hence our focus is on the places where people "do food": where they *eat*, *cook*, and *grow*. We start with consumers (of food, food culture, and food technologies) because changes at the consumer level can retrofit change back up the supply chain of food and bring about large-scale change in food production systems with ramifications throughout food cultures in general. We also start there because HCI interventions often focus in "communicative ecologies" of which the food zone is an exemplar.

Food Zones

Foth and Hearn (2007) introduce the idea of communicative ecologies consisting of three layers: namely, the *material*, the *social*, and the *ideational*. We see food zones in general as consisting of these three layers. The material environment is the most basic layer of a zone, that is, earth, buildings, furniture, roads, and technology infrastructure generally. The social layer consists of the people, their relationships, the organizations, and institutions that are in the zone. Food zones have a symbolic or ideational layer, that is, the stories or meanings that are found or created in the zone. This includes aesthetics and style, the actual and implicit messages in the space, as well as any personal associations the place may possess or invoke.

We suggest three scales of food zones as well. Micro zones are small intimate spaces that are very temporary and circumscribe activity between two or three people. We could think of these as zones of communication or conversation. Meso or mid-range zones encapsulate activities that occupy our attention for several hours. These could include bars, restaurants, playgrounds, community gardens, and, of course, the various spaces and rooms in a house or workplace. The third zone is what we call a macro zone. This is a city-level zone. Most of us live in and relate to our city in highly personal ways. This is not only a matter of access to jobs and services but a personal identification often over extended periods of time. We treat towns and suburbs equally as macro zones.

In evolutionary terms the human body arguably evolved to take part in hunting and gathering, and then farming and agriculture. In the last two hundred years the zones we live in have dramatically transformed how we live our lives in terms of these basic functions. For example, most contemporary citizens are disengaged from the basic processes of food production and preparation. In fact, many of the food production zones that are part and parcel of our Industrial Age cities have had the effect of making food invisible except for the eating of it. This is particularly true in relation to butchering and preparing meat. Virtually all the unpalatable aspects of our food chain are blanked out.

A case can be made, therefore, that macro, meso, and micro food zones are useful frameworks for thinking about how consumers connect with food in their lifestyles. For example, health researchers have coined the term "obesegenic environment" to describe the fact that obesity can be predicted not only from the global geographical zone you live in but also from the city zone as well: "the availability of high energy, dense, palatable, inexpensive food is only surpassed by the mechanized labor-saving and entertainment devices designed to keep us from moving too much" (Ard 2007, 1058). Emerging trends, as exemplified in this book, in the field of more sustainable and healthy food systems include: community gardens; local grower networks; cooking clubs; home delivery services; the resurgence of grower markets and fresh food themed mainstream supermarkets; and greater varieties of traditional and ethnic foods.

Equally, we can think about the role of food meso zones in the consumption process. One of the key elements here is the social component of food zones. Those cultures with strong family bonds or close-knit extended families often have subtle food cultures that mutually help out in managing food-related activities, making them enjoyable, communal, and healthy options. Sharing food with friends facilitates in developing a hospitable and rich food culture which in turn increases the sustainability of this culture. For example, in France, food is understood as part of social life and this builds in healthy food norms. In the nation of Japan, where one of our authors currently resides, it is difficult to imagine a culture more obsessed with all facets of food. Anyone who has flicked through Japan's television channels will not fail to notice the numbers of programs that feature food: eating food, cooking food, quiz shows about food, cooking competitions, food hunters, food fashion, and, more currently (as a result of the earthquake and tsunami of March 11, 2011), commentaries on food availability and safety. It could be construed that in Japan food functions as an interface that smoothes out most interpersonal interactions and exchanges.

Workplaces are equally important food meso zones. How we design our food intake in the meso zone of work is the second place we need consider in creating new food lifestyles.

Finally food micro zones remind us that food can be an intensely intimate and personal experience. The kitchen is a site where not only is food prepared and consumed but equally, it is also inscribed with relational and personal significance. French sociologist Jean-Claude Kaufmann, in *The Meaning of Cooking* (2010, 240), suggests:

Sharing meals with our families weaves the bonds that keep us together. As in the earliest societies, kinship is constructed "by porridge" and by sharing food day after day. . . . The family sat facing one another over a table. Eventually our bodies tired of that soulless rigidity. Our souls for their part were experiencing new desires. They wanted our hearts to beat as one or should that prove impossible, at least to communicate a little in reassuring intimacy. . . . And they wanted to communicate via their senses: food became an existential tool.

The Food Futures Cube

If we combine the three dimensions described above—*food activity* (eat, cook, grow), *scale of zone* (micro, meso, macro), and *layers* (technology, people, meanings)—we have a taxonomy that can then be projected into the future at different time intervals. In theory, we could derive twenty-seven provocations for the future of food, one for each cell of the Food Futures Cube. Here, we present just three. This will not only serve to illustrate the Food Futures Cube (figure 15.1), but it also provide a stimulus for future HCI interventions in the realm of food futures.

Provocation 1: Augmented Reality—Near-Future Storying of Supermarket Foods

In this provocation, we consider four intersecting trends: the growing unrest of food-buying customers; the popularity of community food markets; the trend toward home-based roof-top farming; and the integration of augmented reality (AR) technologies into mainstream mobile information, communications, and media technologies (ICMT). In combination, these four powerful trends starting from the second decade of the twenty-first century forced supermarkets to take extraordinary measures to allay consumer fears over the safety of supermarket foods.

Generations along from primitive electronic food-tagging, AR devices were applied to address consumer concern about supermarket food reliability and traceability. At Costco supermarkets, shoppers picked up not only their baskets at the entrance but also a personalized AR headset. Once donned, the strolling shopper's gaze would activate embedded "story-chips." Detailed biographies of AR-scanned food would appear superimposed over the actual package. Shoppers were able to see and listen to a full life history of the food in question, down to fine details of packaging, how the item

Food futures cube

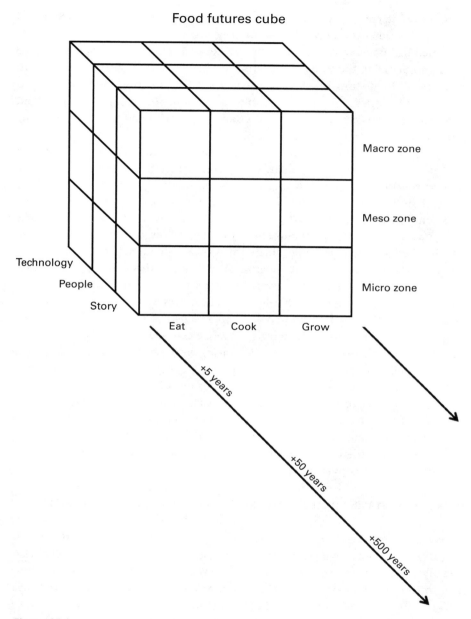

Figure 15.1
The Food Futures Cube

was priced, how it compares to similar products in other retail outlets, and a full breakdown of ingredients and nutritional information.

Augmented reality via bulky VR-type apparatuses and smart phone applications led the way in popularizing the idea of a supermarket-based AR food storying; this transformed everyday food stuffs on the supermarket shelf into living entities with their own stories and histories. Shoppers invariably had become increasingly curious about where their foods were coming from and the stories they each had to tell before being committed to the consumers' dinner table.

These emergent consumer-food-retailer relational developments caused a range of unintended outcomes. For one, this particular supermarket chain transformed into a sort of food tourism spot. Buses would unpack hundreds of food-conscious school teachers and students to wind their way through the markets as though it were a culinary theme park, enthralled by the stories behind the food on every shelf. There were other outcomes: doubting the authenticity of the devices, consumer safety groups would verify the food histories and evaluate them on a systematic "honesty and accuracy scale," a strategy which kept the supermarket administrators and their food storytelling teams constantly vigilant and aware of their scripting quality control.

Furthermore, customer surveys revealed that customers, once barely aware of the origins of the food they bought and cooked, approached their daily bread in unexpected ways. One example is that foods came to be perceived as increasingly animated, and the would-be cook, before preparing each meal, was now inclined to initiate each food item by giving thanks, much in the way that Japanese diners commence each meal with the ritualistic expression *itadakimasu*. This term *itadakimasu* literally means "I receive," an abbreviated form whose meaning is closer to: "I receive this sacrificial meal-as-a-life-form so that I may live." In all, this newfound matrix of values resident in food-consumer dynamics became a rich and diverse form of what Alvin Toffler (1970) termed as an "experiential industry," with as much in common with the global movie industry as with conventional notions of food as a simple means to survival.[1]

Provocation 2: What Will I Print for Daystart (Formerly Known as "Breakfast")?

This provocation explores some "what ifs" of a techno-utopian twenty-first century *Land of Cockaigne* future in which the "personal food printer" (PFP),[2] a magical device that literally prints out meals, became a typical household appliance, and even the names of conventional meals take on a robotic tone (e.g., "breakfast" turns into "daystart"). This is the science-fictional world come true in which diners are able to "print" out any type of food or drink according to a virtually infinite menu of recipes—the ultimate food technology that could eclipse all other cooking technologies. A few "what if" provocations from a PFP future include:

1. *What would happen to the simple pleasure factor of actually cooking food the old-fashioned pre-PFP way?* Perhaps, rather than signifying the end of culinary pleasure, novel and unpredictable forms of pleasure and community spirit would also emerge much as new computer software programs facilitated the democratization of the creation of websites, games, and all sorts of applications, once dominated by highly specialized IT nerds, professionals, and programmers. Perhaps the PFP would function more as a supplement as opposed to a displacer of conventional cooking technologies, expanding the repertoires of the culinarily challenged.

2. *What happens to cooking-generated waste in a PFP world?* Theoretically, such a device has an attractive eco-function in that foodstuffs ought to be theoretically programmable to produce minimal or even zero waste. PFP zero-waste meals consumed in combination with edible eating utensils would coalesce into a utopian environmentally friendly food-cooking and eating system for the future.

3. *Would the PFP spark a neo-Luddite reaction from cooking purists?* We suspect it would but only in part. Like many new technologies, the PFP and similar technologies will spark new combinatory possibilities, new forms, and new ways of performing old activities without entirely displacing preceding technologies. Food printers may not work for all palates and may not translate across all culinary cultures, but they could become one new feature to the mid-twenty-first-century kitchen. The proof will be literally in the pudding: will foods prepared on PFP taste as good as conventionally cooked foods?

4. *How would contemporary restaurants react to the PFP?* We envisage that some commercial food retailers would succumb to the conveniences of PFP-type technologies and self-transform into the culinary equivalents of the Net café with any kind of food available online to print out. PFP cafés could become the next wave of fast food. The term "printer" itself could function as a novel metaphor and effectively help reinvent the arts and sciences of food making as a form of nutritional sculpting—a new art form, with food printing considered as a new category in the creative industries.

5. *Would families bother to even think of shared evening meals when the PFP is on tap twenty-four hours a day?* Techno-pessimists might declare the advent of a home appliance PFP as the end of the family meal, worsening the problem of communication-poor homes. Optimists might hope that such a device would conversely encourage intrafamily communication by functioning like a twenty-first-century fondue dinner set bringing families closer as they delight in its possibilities for collaborative cooking and eating.

6. *Would the PFP, and the variations that would surely follow, eclipse all other kitchen technologies, collapsing the average kitchen into a single "black box" type machine?* Media technologies theorist Henry Jenkins (2006) would be skeptical of the

notion that a new technology such as the PFP would displace current technologies, collapsing all kitchen appliances into a single all-mighty black box. Jenkins refers to this as the black-box fallacy (2006, 26). While the PFP might appear as a utopian technology, we are of the opinion that food printers are more likely to establish their position in the marketplace alongside a diverse range of low-tech to high-tech appliances.

Provocation 3: Legislation of "Neighborhood Food Sanctuaries"

Due to global warming and the ongoing uncertainties of food quality, availability, safety, and security, a worldwide debate on how to rationalize and maximize local community food cultures prompted several UNESCO-recognized Creative Food Cities—small, medium, and mega—to initiate local resolutions calling for the establishment of so-called neighborhood food sanctuaries.[3]

The concept of the neighborhood food sanctuary derived from three sources. First was the legendary science fiction film *Blade Runner*, in which a future society with few real animals made the keeping of at least one artificial animal per household a legal requirement. The second source was from Japan, where some communities had legislated for all neighborhood blocks to be required to have at least one communal park (to co-function as a play area as well as a disaster evacuation zone) and one communal reservoir of water. The third influence was from the philosophy of Austrian-born artist Hundertwasser, who strove to make it mandatory for all buildings to include a roof-top green space to compensate for ground space. Individual rooftop farming grew as a natural extension from this simple idea.

Taking these foundational concepts, the network of Creative Food Cities had taken the lead in realizing the first neighborhood food sanctuaries. To avert locally occurring food crises, each neighborhood block of residences consented in collaborating to build and maintain a plot of land—typically the size of a small park—devoted solely to growing, maintaining, and monitoring local foods.

Due to spatial constraints in high-density residential zones, Buckminster Fuller inspired airborne tensegrity[4] spheres suspended like floating greenhouse balloons above the food sanctuary site and tended by garden carers: local resident gardening aficionados and schoolchildren who participated as part of educational projects. Remote tele-presence technologies meant that local residents could keep watch over their own plants from a distance and use remote virtual and robotic technologies to water, dig, pick ripe foods, and notify block residents of their availability.

There were other features of this micro-community approach to food production. Neighborhood residents would contribute their home food waste to be recycled as compost. This shared community of gardens producing local and seasonal vegetables, herbs, and other nutrient-rich foods added not only a new dimension to the technological aspects of locally grown foods, but also to the community shared meanings of food. Furthermore, like the One World Café[5] type of community cafés popular during the first decade of the twenty-first century, residents were invited and encouraged to take home food at a cost they could afford.

Over the years, novel and diverse "value ecologies" emerged from these neighborhood food sanctuaries. Impromptu gourmet parties would spontaneously happen on the sanctuary site; food researchers and anthropologists would study the sanctuary dynamics hoping to avoid their failures and replicate their successes to other facets of local community life; and creative types would travel far to seek inspiration by immersing themselves within the experiential design-like qualities of these unique spaces. To draw the curtain with one example: one team of soundscape composers attached an array of super-sensitive sensorial devices to plants whose micro-dynamics would be recalculated as algorithms and transformed into real-time music, which could be accessed remotely for connoisseurs of experiential music.

Conclusions

Web 2.0 technologies can be thought of as a way to change our connection to people and meanings in the places we inhabit. Thus they serve as a way for us to think creatively about the different food zones we inhabit. They allow us to experiment with how the different agents and resources in any food zone are related.

As we attempted to make clear from the onset of this chapter, our intentions here were exploratory and diverged considerably from conventional futures approaches to the analysis of futures issues.[6] Our three prototype provocations were intended, of course, to provoke, but provoke what exactly in retrospect?

First, we hoped that the Food Futures Cube with its micro/meso/macro layering would provoke a more layered and interconnected approach to thinking about food cultures in general, and more importantly, by introducing this strange yet familiar mathematical structure, novel and usually hidden properties pertaining to food would emerge and make themselves visible. Indeed we have not suggested the cube as the ultimate shape from which to consider food futures, and in fact, our readers might be provoked to suggest their own frameworks: Buckminster Fuller would have relished such a challenge.

Second, we hoped our provocations will provoke the reader to use these tentative stories as a platform to build upon (or take away from if the story is morally distasteful), and as a resource from which to extract points and ideas worthy of further analysis. If we have provoked our readers to forge (or even just think about initiating) novel ways of approaching food cultures, then we have been successful.

DAVE AND GREG'S TRAVELING CURRY

Makes 5–8 servings

Our recipe is a curry that exemplifies a story of diaspora and culinary evolution. It is also extremely delicious. The curry is like a jazz improvisation: a basic structure but the riffs are slightly different every time it is performed. Many of our curry fans have asked us for the recipe and reacted with suspicion when we told them we were not really sure how to make it. This is the first time we have quantified this recipe for replication in someone else's kitchen.

This is also a recipe that personally brings relief from academic stress—cooking as therapy. This curry has subsequently taken on a life, indeed, many lives of its own as it is told and retold, as a form of emergent cooking. Although our curry was inspired in the islands of the Seychelles by the variety herbs and spices available there, our first encounter with it in the kitchen was in 2000, in Brisbane, Australia. The traveling curry continues to travel and evolve. We recommend the traveling curry be not eaten alone but in the company of fellow travelers and food raconteurs. Serve and decorate on the plate with a splash of yogurt and adorn with herb leaves of choice (ours are different every time depending on the season). Accompany the traveling curry with rice and full-bodied red wine for maximum creative culinary experience.

Ingredients

2 medium onions, diced

4–5 garlic cloves, chopped

½ cup minced ginger root

1 tbsp. sesame seeds

6–10 dried chili peppers

5–8 bay leaves

Salt and pepper to taste

Olive oil

Herbs and spices such cinnamon, allspice, turmeric, rosemary, 3–4 star anise (ground), and any others you like (we call our mixture "Seychelles curry powder")

Soy sauce and honey to taste

1 cup medium-bodied red wine

½ cup mango, chopped or pureed (if mango is not available, blueberries or persimmons are a good substitute for a fruity-spicy balance)

6–8 whole tomatoes, chopped

2 cups coconut milk

10–12 raw chicken thighs

sliced lotus root (optional, for crunch)

2 medium eggplant, sliced

2 cups sliced mushrooms, mixing variety, such as shiitake, oyster, or cremini, if desired

10–15 pitted black olives, halved

8–10 hard-boiled quail eggs

Preparation

1. Cook onions, garlic, and ginger over a low flame in a large pot until onions begin to soften. Add sesame seeds, chili peppers, bay leaves, salt and pepper.

2. Add some olive oil—as much as you need to keep the mixture from burning as it continues to lightly brown.

4. After the onion is translucent, add the curry powder. When the contents of the pot start to condense into a paste, add a bit of soy sauce, the wine, and honey.

5. After a few minutes add the mango or fruit pieces, and the tomatoes.

6. Add coconut milk and stir until the sauce is evenly colored.

7. Place chicken thighs in pot, letting them settle into the mixture. Slowly allow to simmer for at least 30 minutes until tender.

8. Add eggplant, mushrooms, and olives. Cook until eggplant is tender, about 20–25 minutes.

9. Once the curry is more or less ready to be served, place the quail eggs on top and gently mix in so as not to squash them. Kanakis sauce or prune juice can also be used as a fruity flavoring if you need more liquid, but it might clash with the mango. Let your taste buds make the call. We have also been known to top off the dish with either a few cubes of grated dark chocolate and/or finely ground coffee—at the risk of overcomplicating the subtle flavors.

Notes

1. User-Generated Contents proof-of-concept promotion video to illustrate the AR device embedded in supermarket environments and its implications accessed from www.futures-tv.net—the dedicated futures issues broadband TV channel.

2. The PFP presented here is based on an actual device, the Personal Food Factory, currently in development by researchers Marcelo Coelho and Amit Zoran of the Fluid Interfaces Group at MIT.

3. Source: UNESCO Creative Food Cities Network News, 2022.

4. Fuller coined the term "Cloud Nine" to refer to these airborne habitats created from geodesic spheres, which if heated slightly would rise into the air. The term "tensegrity" is Fuller's contraction of "tensional integrity."

5. One World Café is a nonprofit community café based in Salt Lake City, Utah. OWC operates on the so-called pay-what-you-want model, as there are no prices or menus at the café.

6. The early sections draw on David Wright (2010), Unpacking Japan's 21st Century "National Conversation": Images of the Future beyond the Iron Cage of the "Catch Up" Model. Unpublished doctoral thesis, Queensland University of Technology.

References

Ard, J. D. 2007. Unique perspectives on the obesegenic environment. *Journal of General Internal Medicine* 22 (7):1058–1060.

Boulding, Elise. 1988. Image and action in peace making. *Journal of Social Issues* 44 (2):17–37.

Carey, John, ed. 1999. *The Faber book of utopias*. London: Faber and Faber.

Coleman, Stephen, and Paddy O'Sullivan, eds. 1990. *William Morris and* News from Nowhere: *A vision for our time*. Bideford: Green Books.

Foth, Marcus, and Gregory N. 2007. Networked individualism of urban residents: discovering the communicative ecology in inner-city apartment buildings. *Information, Communication & Society* 10 (5):749–772.

Gorman, Michael John. 2005. *Buckminster Fuller: Designing for mobility*. Milano: Skira.

Inayatullah, Sohail. 2003. Teaching futures studies: From strategy to transformative change. *Journal of Futures Studies* 7 (3):35–40.

Jenkins, Henry. 2006. *Convergence culture: Where old and new media collide*. New York: New York University Press.

Kaufmann, Jean-Claude. 2010. *The meaning of cooking*. Cambridge: Polity Press.

Leadbeater, Charles. 2002. *Up the down escalator: Why the global pessimists are wrong*. London: Penguin Books.

Pleij, Herman. 2001. *Dreaming of Cockaigne: Medieval fantasies of the perfect life*. New York: Columbia University Press.

Polak, Fred. 1973. *The image of the future*. Boulding, Elise, (trans.). New York: Elsevier.

Russell, Bertrand. 1952. *The impact of science on society*. London: Allen & Unwin.

Smith, Patrick. 1998. *Japan: A reinterpretation*. New York: Vintage Books.

Tsuji, Shinichi. 2001. *Slow is beautiful: Slowness as culture*. Tokyo: Heibonsha.

Toffler, Alvin. 1970. *Future shock*. New York: Random House.

Epilogue: Bringing Technology to the Dining Table

Charles Spence

While technology and food might initially seem like rather strange bedfellows, the various chapters in this volume highlight a number of ways in which digital technologies increasingly influence our everyday experience with food: everything from how we source, find out more about, and keep track of our diminishing food supplies (see chapters 1 and 7) to the way in which we consume and/or interact with food and drink (see chapters 4 and 6). The emphasis in this volume is very much on the question of how new digital technologies may come to change the landscape of eating (and, in some cases, how they already have). This fascinating area of research, exemplified by the list of authors who have been brought together within the preceding pages, bridges the divide between academia and industry (e.g., see the intriguing contribution from Philips Research in chapter 6).

My own interest in this area centers on the question of how emerging technologies will increasingly come to be integrated into, and hence change (hopefully for the better), our dining experiences, whether we are sitting around the dinner table with friends and family (see chapter 4 and also Barden et al. 2012), or dining out at one of the increasing popular restaurants serving molecular (or modernist) cuisine (e.g., Blumenthal 2008; Myhrvold and Young 2011). While the tremendous growth of modernist cooking in recent years has relied, at least in part, on the development and utilization of new technologies in the kitchen (Vega et al. 2012), there seems to be tremendous scope here to revolutionize our eating and drinking experiences/behaviors through the intelligent marriage of food and drink with the latest in digital technology, be that at the point of purchase (see chapter 7), at the point of creation (i.e., when preparing or cooking food; see chapter 6), at the point of consumption (see the chapters in the first section, "Eat"), or even in the act of growing the raw ingredients themselves (see the chapters in the final section, "Grow").

This trend is already taking hold in some places. A number of restaurateurs have already started to experiment with the projection of images onto the dinner table[1] or to allow diners to change the color of the lighting in their dining space.[2] (At Inamo, a recently opened restaurant in London,[3] diners can place their orders from an illustrated food and drinks menu projected on to their table surface. As the restaurant owners state on their website: "At the core of Inamo is our interactive ordering system. . . . You'll set the mood, discover the local neighbourhood, and even order a taxi home."

Dishes such as "Sound of the Sea" (featured on the tasting menu at Heston Blumenthal's The Fat Duck restaurant in Bray, Berkshire, United Kingdom)[4] highlight how technology can be used to deliver a genuinely different kind of dining experience (Spence, Shankar, and Blumenthal 2011). The waiter arrives at the table holding in one hand a plate of seafood covered in seafood foam resting on a "beach" made from tapioca, breadcrumbs, and sashimi. With his other he presents a seashell out of which dangles a pair of iPod earbuds (see figure 16.1). The waiter recommends that the diner put the earbuds on before starting to eat. What the diner hears, assuming he or she does as told, is literally the sound of the sea: the waves crashing gently on the beach, a few seagulls flying overhead. Some diners have been known to find the multisensory experience so powerful that they have broken into tears (de Lange 2012).

In the case of "Sound of the Sea," the technology (nothing more than a iPod mini) completely transforms the dining experience, both by enhancing the taste of the dish (Spence and Shanker 2010), and by getting the diner to pay more attention to the gustatory experience. When last I dined at Blumenthal's flagship restaurant in the

Figure 16.1
The Sound of the Sea seafood dish

small village of Bray, it was striking how nearby tables of erstwhile talkative diners suddenly became mute once they had put the earbuds into their ears. In part, the idea is for the diner to come away from a meal at The Fat Duck thinking rather more carefully about the multisensory dining experience, and how what we hear plays a role in our experience of what it is we eat and drink. As Heston himself puts it: "Sound is one of the ingredients that the chef has at his/her disposal" (cited in Spence 2011). For me, the excitement of merging digital technology with food and dining lies precisely in its potential to radically change many of our dining experiences, and to do so for the better. I believe this potential will exist for a Michelin-starred molecular gastronomy chef or someone getting ready to host a dinner party at home. In fact, I would argue that a number of the most successful advances (and technologies) emerging from the world's top molecular gastronomy kitchens today will likely make their way into some homes within a few years.

In the proceedings of many an international conference on human-computer interaction, I see an increasing number of food-related augmented reality (AR) applications now being developed, and gaining a great deal of publicity in the media. Many of these technologies enable users to experience food in unique ways: Take, for example, the EverCrisp app, developed by Kayac Inc. of Japan (but, as yet, not licensed by Apple) (see figure 16.2).

The idea with EverCrisp was to use a mobile device to enhance the crunch of noisy (e.g., dry) food products simply by changing the sound that people heard as they bit into the food.[5] Other researchers, meanwhile, are currently working on technologies, such as the "mouth jockey," which will allow people to synchronize a variety of different sounds whenever they close their jaw while eating (Koizumi et al. 2011) (see figure 16.3).

A few years ago Yuki Hashimoto and colleagues (2007; 2008), also working out of Japan, developed a straw-like user interface. This AR device re-creates the sounds and feeling associated with sucking a particular liquidized (or mashed) food through a straw: simply place the straw over a mat showing a desired food and then suck on the straw. Having experimented with this interface myself, I can confirm how realistic the audiotactile experience is, even though no actual food passed my lips. At the same time, many other research groups (in Japan and elsewhere) are currently working on AR and virtual reality (VR) applications to enable us to change the apparent color, texture, and even the size of the food we are eating (e.g., Narumi et al. 2010; Okajima and Spence 2011; Sakai 2011).[6]

Although such technological innovations (see figure 16.4) certainly help to highlight the possibilities of a marriage between technology and food, I am concerned that, as yet, many of those working in the HCI arena focus too much energy on showcasing what the technology can deliver without spending enough time thinking about the practicalities associated with using the technology in the restaurant or home-dining setting.

Figure 16.2
EverCrisp app

However, over-and-above any potential use of technology to enhance our experience of food and drink, more and more researchers are investigating whether digital technologies can also be intelligently used to help control our eating behaviors. Undoubtedly the worldwide obesity crisis represents one of the more serious challenges facing society today (e.g., Marteau, Hollands, and Fletcher 2012). In chapter 6, Toet and colleagues report on the potential use of digital cutlery and plate ware. Their idea is that technology may help monitor how much people are eating, and even, potentially, "nudge" a diner (Thaler and Sunstein 2008) to eat more slowly, or perhaps to cease eating entirely once the technology has decided a sufficient (or a predetermined number of) calories has been consumed. While such research on the use of digital technology to improve our eating behaviors is still in its infancy, it nevertheless appears to represent a promising, not to mention important, area for future research.

Another problem in our disconnected society relates to the fact that many of us often find ourselves working in a different city or even country than our family. As a result, people often miss out on shared family time, which is often focused around the dinner table. In chapter 4, Comber and colleagues report on the use of technology to

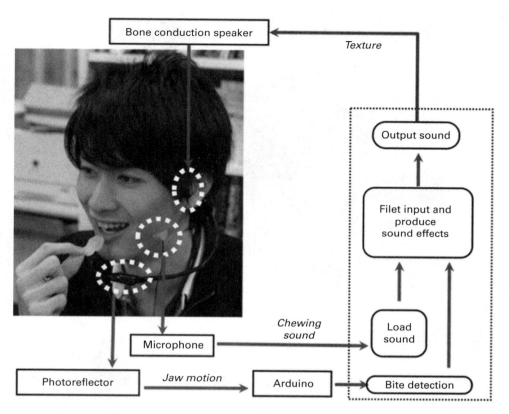

Figure 16.3
Mouth Jockey (Koizumi et al. 2011)

allow those situated far apart to share meaningful virtual mealtime/dining experi-ences—what has been described as "the telematic dinner party" (see Barden et al. 2012). Although the authors themselves freely admit that further research is needed on this topic before any workable HCI solution is likely to be available, the findings that have emerged to date are nevertheless intriguing.

Today a quick scan of cutting-edge restaurants shows a recent increase in the use of technology at the table; I have already mentioned restaurants where the diner can interact with technology at the table. Although the use of such HCI at the table may well provide a means of distraction,[7] or of differentiating oneself from the opposition in the challenging world of fine dining, I suspect that such technologies will only stick in the marketplace if they benefit or enhance the diner's experience. Indeed, many diners are sensitive to the danger that the multisensory dining experience has, in certain cases, already become more important than the food/drink itself (Gill 2007;

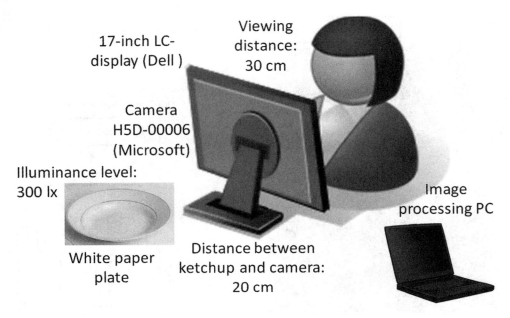

Figure 16.4
Augmented reality food (Okajima and Spence 2011)

Goldstein 2005; Spence and Piqueras-Fiszman 2012), and there is always going to be a danger that the food will suffer as a result whenever technology takes center stage. So, for example, one could think of adding real value to the dining experience at a restaurant where the diner is put in control of the ambient lighting by marrying the technology with research findings demonstrating that wine (and presumably other food and beverage products) taste sweeter when consumed under red ambient lighting (Oberfeld, Hecht, Allendorf, and Wickelmaier 2009). Suddenly, the technology conveys a meaningful benefit (over-and-above any entertainment value it may have); it allows for one to make the food/drink taste sweeter without the need to reach for the sugar. Similarly, other research suggests that the overall lighting level may affect how much you will enjoy your cup of coffee (Gal, Wheeler, and Shiv 2007). Again, HCI could be put at the service of enhancing the diner's experience rather than merely acting as an entertaining distraction.

However, over and above any use of technology to change the visual attributes of the dining environment, a more interesting use of technology may be to match the auditory atmosphere to the food/drink we happen to be consuming. A large body of research now shows the profound effect of what we listen to on everything from the food and drink choices we make through to the experience of the taste of the food and drink itself (Crisinel et al. 2012; North 2012; Spence 2012a; Spence and Shankar 2010). In the future, we may see an increased use of technology allowing for the personalized delivery of music and/or soundscapes to individual tables (where a group of friends may be sharing a bottle of wine), or even to an individual diner or drinker. The last few years have already seen something of an explosion of research on the matching of music/soundscapes to the flavors of specific foods and drinks (Crisinel et al. 2012; Knöferle and Spence 2012; Mesz et al. 2012; Spence 2011b). When such insights are paired with the recent developments in musical plateware—cups and plates that make music whenever they are picked-up, or which change the sound they make as the level in your glass slowly goes down[8]—who knows what fun might be had. The use of hyperdirectional loudspeakers[9] may also provide for some intriguing opportunities for the targeted delivery of experiential soundscapes for drinkers/diners.

One of the challenges we are working on currently at the Crossmodal Research Laboratory here in Oxford goes by the name of "Gin and sonic." Denis Martin, a chef with two Michelin stars who runs the restaurant Denis Martin in Vevey, Switzerland,[10] uses a balloon and a liquid nitrogen bath to create a gin and tonic the likes of which hasn't been seen before (see figure 16.5). While the dish is visually dramatic, the one thing missing is the sound of the tonic gently fizzing in the glass. We are currently

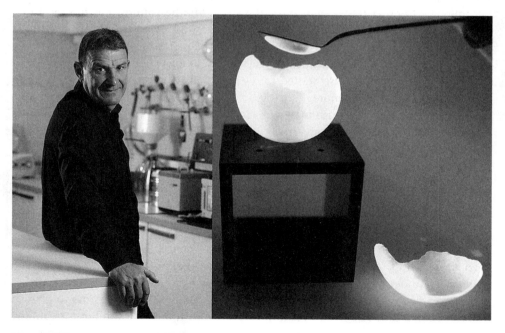

Figure 16.5
Gin and sonic

working with Condiment Junkie on embedding actuators into the plateware in order to bring back the sounds of carbonation (Zampini and Spence 2005).

Furthermore, marketers are just now starting to come out with musical apps to support/enhance their customer's brand experiences.[11] I predict that digital technology will increasingly be used to support the explosion in synaesthetic marketing and multisensory stimulation that has been forecast for a number of years (Meehan et al. 1998) and finally seems to be upon us (Spence 2012b). Meanwhile, at the opposite extreme, other researchers are thinking about how to harness emerging technologies in order to reduce the din than many of us complain about in restaurants (Clynes 2012).

No discussion of technology and food, however, would be complete without thinking about the Internet. Given all of the information at our disposal over the web, is it not strange that we mostly leave the decision of what to order from the menu until we actually arrive at the restaurant itself (a time surely that we would rather spend chatting with our dining companions)? In chapter 3, Jettie Hoonhout and colleagues investigate the possibilities (and challenges) associated with the digitization of menu/recipe recommendations, and how technology could potentially be used to support healthy food choices (e.g., by providing us with the relevant nutritional information at the most appropriate time). Surely I can't be the only one wondering how long it will take before the dominant place to look for a new recipe (or even an old favorite) becomes the Internet rather than those dog-eared and bespattered cookbooks sitting forlornly on the kitchen shelf. As of 2011, Google was already reporting that more than 1 percent of the queries it received were for recipes; no wonder, then, that they recently introduced a recipe feature (Singel 2011). Famous chefs such as Jamie Oliver and Nigella Lawson have also started to release their own apps for those wanting to learn more about (and do more) cooking.[12]

Digital technology may also play a crucial role in helping to document the many recipes and innovations that come streaming out of the kitchens of today's top-ranking chefs, especially from those cutting-edge molecular gastronomy chefs who are forever searching for the next surprise to deliver to the dining table (Piqueras-Fiszman and Spence 2012). Indeed, since closing his world famous restaurant, ElBulli, in Spain, Ferran Adrià has devoted some of his energies to creating a digital library capable of documenting and preserving all the wealth of culinary creations that have emerged from his kitchen over the years (Williams 2012).

While many people intuitively dislike the idea of mixing technology with food, even those who support the Slow Food movement (Petrini 2007), may be forced to embrace technology in order to achieve the cultural movement's stated philosophical objectives (see chapter 5). As stated earlier in this volume, HCI may come to play an especially important role in connecting consumers with producers in both urban and "rur-urban" areas (Donadieu 2005). Mobile and social technologies have the potential to facilitate our access to information about food at the point of purchase and decision making, to connect with local food producers, and to share our food experiences and knowledge with our peers and online acquaintances (e.g., "followers" or "friends"). Several chapters discuss the use of digital technology to connect food communities: for example, in chapter 2, Kit MacFarlane and Jean Duruz report on an anthropological study documenting one successful example of targeting restaurant customers in Australia via online social networking (Facebook). The technology in this case, once again, was being used to promote, establish, and sustain a food culture/community in which it might be difficult to do so by other means.

I end with one of my current favorite ideas on the theme of digital technology at the dining table: just think for a moment about how the eating experience could be changed/enhanced if people were to stop being so distracted by their tablets (and other handheld mobile devices) while eating. Thus my radical suggestion would be to ask, why don't we turn our tablets into plates (see figure 16.6)?

What possibilities might open up if we were to start serving food from a tablet? One idea that immediately springs to mind is to develop the ability change the screen color and hence the plate color in order to bring out the sweetness in a dish, for example. This suggestion is based on recent findings showing that a strawberry dessert was rated as tasting 10 percent sweeter, and 15 percent more flavorful when eaten from a white plate rather than a black plate (Piqueras-Fiszman et al. 2012; Harrar, Piqueras-Fiszman, and Spence 2011). One might also be able to trigger particular kinds of music (or soundscape) depending on what exactly the diner chooses to eat from the plate. I am also looking forward, in the years to come, to the use that artists/conceptual chefs will undoubtedly and increasingly make of the emerging digital technologies when

Figure 16.6
Tablets into plates

working with food (Jones 2006; Marinetti 1930/1989). And if one thinks that dining from a tablet sounds bizarre, one need only mention the chicken liver parfait that was being served from a brick in one of London's hottest new restaurants in 2012 (John Salt)[13] for comparison (London Tastin' 2012).

Conclusions

I hope after reading the array of chapters in this intriguing volume that you will share my enthusiasm and excitement when it comes to thinking about the various ways in which digital technologies will increasingly change our interaction with food and drink in the coming years (see Spence and Piqueras-Fiszman in press, for further discussion of this topic). As we saw at several points throughout this book, the effective marrying of food with technology can prove challenging. Nevertheless, I am convinced that the potential benefits will undoubtedly be worth the effort. So let's move forward and really start bringing the technology to the table.

Notes

1. See, for example, http://vimeopro.com/user10658925/el-celler-de-can-roca/video/40919096.

2. See, for example, Pod, Philadelphia, http://www.podrestaurant.com.

3. See http://www.inamo-restaurant.com/pc.

4. See http://www.thefatduck.co.uk.

5. See Spence 2012a and Zampini and Spence 2004, for the background science.

6. See also http://www.youtube.com/watch?v=cbsKSPdOSX0; http://www.newlaunches.com/archives/one_cookie_seven_flavors_with_meta_cookie_ar_headgear.php; and http://www.springwise.com/food_beverage/qr-code-cookies-create-unique-personalized-messages.

7. See Bellisle and Dalix 2001 on the dangers of distraction for food consumption.

8. See, for example, the musical coffee player designed by Dr. Ju-Hwan Lee, at http://www.youtube.com/user/virsomio?nomobile=1 and http://interactionlab.kr/Project-Contents_1-5.html.

9. See http://dakotaaudio.com/?gclid=CKj8-72rrLQCFe7MtAodpy4Agg and http://www.holosonics.com/technology.html.

10. http://www.denismartin.ch.

11. See, for example, Le Nez de Courvoisier App, http://courvoisier.com/uk/le-nez-de-courvoisier-app.

12. See http://www.jamieoliver.com/apps.

13. See http://john-salt.com.

References

Barden, Pollie, Rob Comber, David Green, Daniel Jackson, Cassim Ladha, Tom Bartindale, Nick Bryan-Kinns, Tony Stockman, and Patrick Olivier. 2012. Telematic dinner party: Designing for togetherness through play and performance. In *Proceedings of ACM Conference on Designing Interactive Systems 2012 (DIS2012)*, 38–47. New York: ACM.

Bellisle, France, and Anne-Marie Dalix. 2001. Cognitive restraint can be offset by distraction, leading to increased mean intake in women. *American Journal of Clinical Nutrition* 74:197–200.

Blumenthal, Heston. 2008. *The big Fat Duck cookbook*. London: Bloomsbury.

Clynes, Tom. 2012. A restaurant with adjustable acoustics. *Popular Science*. http://www.popsci.com/technology/article/2012-08/restaurant-adjustable-acoustics.

Crisinel, Anne-Sylvie, Stefan Cosser, Scott King, Russ Jones, James Petrie, and Charles Spence. 2012. A bittersweet symphony: Systematically modulating the taste of food by changing the sonic properties of the soundtrack playing in the background. *Food Quality and Preference* 24: 201–204.

Crisinel, Anne-Sylvie, Caroline Jacquier, Ophelia Deroy, and Charles Spence. 2013. Composing with cross-modal correspondences: Music and smells in concert. *Chemosensory Perception* 6: 45–52.

de Lange, Catherine. 2012. Feast for the senses: Cook up a master dish. *New Scientist*, 2896 (December 18). http://www.newscientist.com/article/mg21628962.200-feast-for-the-senses-cook-up-a-master-dish.html.

Donadieu, Pierre. 2005. *Campaignes urbaines*. Arles, France: Actes Sud, Ecole nationale superieure du paysage.

Gal, David. Wheeler, S. C., and Shiv, B. 2007, unpublished manuscript. *Cross-modal influences on gustatory perception*. http://ssrn.com/abstract=1030197.

Gill, A. A. 2007. *Table talk: Sweet and sour, salt and bitter*. London: Weidenfeld & Nicolson.

Goldstein, Darra. 2005. The play's the thing: Dining out in the new Russia. In *The taste culture reader: Experiencing food and drink*, ed. C. Korsmeyer, 359–371. Oxford: Berg.

Harrar, Vanessa, Betina Piqueras-Fiszman, and Charles Spence. 2011. There's no taste in a white bowl. *Perception* 40:880–892.

Hashimoto, Yuki, Naohisa Nagaya, Minoru Kojima, Satoru Miyajima, Junichiro Ohtaki, Akio Yamamoto, Tomoyasu Mitani, and Masahiko Inami. 2007. Straw-like user interface: Virtual experience of the sensation of drinking using a straw. In *Proceedings World Haptics 2007*, 557–558. Los Alamitos, CA: IEEE Computer Society.

Hashimoto, Yuki, Masahiko Inami, and Hiroyuki Kajimoto. 2008. Straw-like user interface (II): A new method of presenting auditory sensations for a more natural experience. In *Eurohaptics 2008, LNCS*, ed. M. Ferre, 484–493. Berlin: Springer-Verlag.

Jones, Caroline, ed. 2006. *Sensorium: Embodied experience, technology, and contemporary art.* Cambridge, MA: MIT Press.

Knöferle, Klemens, and Charles Spence. 2012. Crossmodal correspondences between sounds and tastes. *Psychonomic Bulletin & Review* 19:992–1006.

Koizumi, Naoya, Hidekazu Tanaka, Yuji Uema, and Masahiko Inami. 2011. Chewing jockey: Augmented Food Texture by using sound based on the cross-modal effect. In *Proceedings of ACE'11, Proceedings of the 8th International Conference on Advances in Computer Entertainment Technology*, Article No. 21. New York: ACM.

London Tastin'. 2012. John Salt: Spectacular culinary experience. *London Tastin'.* http://london-tastin.com/post/36500907797/john-salt-spectacular-culinary-experience.

Marinetti, Filippo. 1930/1989. *The futurist cookbook.* Brill, S., (trans.). San Francisco: Bedford Arts.

Marteau, Thersa, Gareth Hollands, and Paul Fletcher. 2012. Changing human behaviour to prevent disease: The importance of targeting automatic processes. *Science* 337:1492–1495.

Meehan, Mary, Larry Samuel, and Vickie Abrahamson. 1998. *The future ain't what it used to be: The 40 cultural trends transforming your job, your life, your world.* New York: Riverhead Books.

Mesz, Bruno, Mariano Sigman, and Marcos Trevisan. 2012. A composition algorithm based on crossmodal taste-music correspondences. *Frontiers in Human Neuroscience* 6 (71):1–6.

Myhrvold, Nathan, and Chris Young. 2011. *Modernist cuisine. The art and science of cooking.* USA: Ingram Publisher Services.

Narumi, Takuji, Takashi Kajinami, Tomohiro Tanikawa, and Michitaka Hirose. 2010. *Meta cookie. SIGGRAPH '10 ACM SIGGRAPH 2010 Emerging Technologies*, Article No. 18. New York, NY: ACM.

North, Adrian. 2012. The effect of background music on the taste of wine. *British Journal of Psychology* 103:293–301.

Oberfeld, Daniel, Heiko Hecht, Ulrich Allendorf, and Florian Wickelmaier. 2009. Ambient lighting modifies the flavor of wine. *Journal of Sensory Studies* 24:797–832.

Okajima, Katsunori, and Charles Spence. 2011. Effects of visual food texture on taste perception. *Perception* 2 (8). http://i-perception.perceptionweb.com/journal/I/article/ic966.

Petrini, Carlo. 2007. *Slow food: The case for taste.* McCuaig, W., (trans.). New York: Columbia University Press.

Piqueras-Fiszman, Betina, Jorge Alcaide, Elena Roura, and Charles Spence. 2012. Is it the plate or is it the food? Assessing the influence of the color (black or white) and shape of the plate on the perception of the food placed on it. *Food Quality and Preference* 24:205–208.

Piqueras-Fiszman, Betina, and Charles Spence. 2012. Sensory incongruity in the food and beverage sector: Art, science, and commercialization. *Petits Propos Culinaires* 95:74–118.

Sakai, Nobuyuki. 2011. Tasting with eyes. *Perception* 2 (8). http://i-perception.perceptionweb.com/journal/I/article/ic945.

Singel, Ryan. 2011a. Google recipe search cooks up next gen of search. *Wired*, February 24. http://www.wired.com/business/2011/02/google-recipe-semantic.

Spence, Charles. 2011b. Sound design: How understanding the brain of the consumer can enhance auditory and multisensory product/brand development. In *Audio Branding Congress Proceedings 2010,* eds. Kai Bronner, R. Hirt, and C. Ringe, pp. 35–49. Baden-Baden, Germany: Nomos Verlag.

Spence, Charles. 2011. Wine and music. *The World of Fine Wine* 31: 96–104.

Spence, Charles. 2012a. Auditory contributions to flavour perception and feeding behaviour. *Physiology & Behavior* 107:505–515.

Spence, Charles. 2012b. Synaesthetic marketing: Cross sensory selling that exploits unusual neural cues is finally coming of age. *Wired World in* 2013 (November):104–107.

Spence, Charles, and Betina Piqueras-Fiszman. 2012. Dining in the dark: Why, exactly, is the experience so popular? *Psychologist* 25:888–891.

Spence, Charles, and Betina Piqueras-Fiszman. In press. *The perfect meal.* Oxford: Wiley.

Spence, Charles, and Maya Shankar. 2010. The influence of auditory cues on the perception of, and responses to, food and drink. *Journal of Sensory Studies* 25:406–430.

Spence, Charles, Maya Shankar, and Heston Blumenthal. 2011. "Sound bites": Auditory contributions to the perception and consumption of food and drink. In *Art and the senses*, ed. F. Bacci and D. Melcher, 207–238. Oxford: Oxford University Press.

Thaler, Richard, and Cass Sunstein. 2008. *Nudge: Improving decisions about health, wealth and happiness.* London: Penguin.

Vega, César, Job Ubbink, and Erik van der Linden, eds. 2012. *The kitchen as laboratory: Reflections on the science of food and cooking.* New York: Columbia University Press.

Williams, Greg. 2012. After elBulli: Ferran Adrià on his desire to bring innovation to all. *Wired*, September 24. http://www.wired.co.uk/magazine/archive/2012/10/features/staying-creative-ferran-adri%C3%A0?page=all.

Zampini, Massimiliano, and Charles Spence. 2004. The role of auditory cues in modulating the perceived crispness and staleness of potato chips. *Journal of Sensory Science* 19:347–363.

Zampini, Massimiliano, and Charles Spence. 2005. Modifying the multisensory perception of a carbonated beverage using auditory cues. *Food Quality and Preference* 16:632–641.

List of Recipes

Note: The recipes in this book have not undergone rigorous kitchen testing, and are intended as illustrations of the "eat, cook, grow" theme.

Index